LONDON MATHEMATICAL SOCIETY LECTURE NOTE SERIES

Managing Editor: Professor J.W.S. Cassels, Department of Pure Mathematics and Mathematical Statistics, University of Cambridge, 16 Mill Lane, Cambridge CB2 1SB, England

The titles below are available from booksellers, or, in case of difficulty, from Cambridge University Press.

London Mathematical Society Lecture Note Series. 238

Representation Theory and Algebraic Geometry

Edited by

A. Martsinkovsky
Northeastern University

G. Todorov
Northeastern University

CAMBRIDGE
UNIVERSITY PRESS

PUBLISHED BY THE PRESS SYNDICATE OF THE UNIVERSITY OF CAMBRIDGE
The Pitt Building, Trumpington Street, Cambridge CB2 1RP, United Kingdom

CAMBRIDGE UNIVERSITY PRESS
The Edinburgh Building, Cambridge, CB2 2RU, United Kingdom
40 West 20th Street, New York, NY 10011-4211, USA
10 Stamford Road, Oakleigh, Melbourne 3166, Australia

First published 1997

A catalogue record for this book is available from the British Library

ISBN   0 521 57789 6 paperback

Transferred to digital printing 2002

Contents

Preface

Maurice Auslander passed away on November 18, 1994, in Trondheim, Norway at the age of sixty-eight. A memorial conference was held in his honor March 24-26, 1995 at Brandeis University. Over seventy people from all over the world attended the conference to show their respect for this remarkable mathematician and for his accomplishments. This feeling was shared by many others.

The variety of topics covered at the conference reflects the breadth of Maurice Auslander's contribution to mathematics, which includes commutative algebra and algebraic geometry, homological algebra and representation theory. He was one of the founding fathers of homological ring theory and representation theory of Artin algebras. Undoubtedly, the most characteristic feature of his mathematics was the profound use of homological and functorial techniques.

The Memorial Service was held on March 24. Appreciation of various sides of Maurice's personality could be felt throughout the morning of shared memories. From the kind man as seen by one little boy to the unapproachable man in the eyes of others, Maurice had left impressions in the lives of many. Initial fears by students later turned into lasting friendships. Those famous early morning phone calls, his sudden appearances at the offices of students and colleagues, long walks in many cities and countries, all became nostalgic memories. Yet his thoughtful comments and questions, and his many words of advice are still present among those who knew him.

Perhaps it can be said that Maurice's mathematics was a continuation of his personality. Both were characterized by integrity, intellectual honesty, and an everlasting persistence. Even the illness did not stop him. He insisted on enjoying the life, friends, and mathematics until the very last days. He will be remembered as a man who never gave up.

A. M.
G. T.
August 9, 1996
Geiranger, Norway

SOME PROBLEMS ON
THREE-DIMENSIONAL GRADED DOMAINS

M. ARTIN

1. Introduction.

One of the important motivating problems for ring theory is to describe the rings which have some of the properties of commutative rings. In this talk we consider this problem for graded domains of dimension 3. The conjectures we present are based on ideas of my friends, especially of Toby Stafford, Michel Van den Bergh, and James Zhang. However, they may not be willing to risk making them, because only fragments of a theory exist at present. Everything here should be taken with a grain of salt. I am especially indebted to Toby Stafford for showing me some rings constructed from differential operators which I had overlooked in earlier versions of this manuscript.

To simplify our statements, we assume throughout that the ground field k is algebraically closed and of characteristic zero, and that our graded domain A is generated by finitely many elements of degree 1. The properties which we single out are:

1.1.
 (i) *A is noetherian,*
 (ii) *there is a dualizing complex ω for A such that the Auslander conditions hold, and*
 (iii) *the Gelfand-Kirillov dimension behaves as predicted by commutative algebra.*

A *dualizing complex* ω is a complex of bimodules such that the functor $M \mapsto M^D = \underline{\mathrm{RHom}}(M, \omega)$ defines a duality between the derived categories of bounded complexes of finite left and right A-modules. We will require it to be *balanced* in the sense of Yekutieli, which means that k^D is the appropiate shift of k (see [Aj,Y1,Y2] for the precise definitions). A graded ring A with a dualizing complex satisfies the *Auslander conditions* if for any finite A-module M and any submodule $N \subset \mathrm{Ext}^q(M, \omega)$, $\mathrm{Ext}^p(N, \omega) = 0$ for $p < q$ [Bj,Le,Y1,Y2,ASZ].

Typeset by $\mathcal{A}\mathcal{M}\mathcal{S}$-TeX

Let

$$j(M) = min\{j \mid \operatorname{Ext}^j(M, \omega) \neq 0)\}.$$

For domains of dimension 3, the link with Gelfand-Kirillov dimension is that $gk(M) = 3 - j(M)$ [Le,Ye2,YZ]. Actually, the GK-dimension is not always the right dimension to use (see [ASZ]), but it will suffice for our purposes.

Though the properties listed above are central, they will appear only implicitly in what follows. All of them hold when A is commutative. So our problem becomes: Which graded domains satisfy these conditions? An answer to this question might take the form of an axiomatic description, or of a classification. This talk concerns classification, for which I should apologize. Maurice looked askance at what he might have called "botany", so the topic is not very suitable for the Auslander Conference.

Let's begin by reviewing the commutative case. If a commutative graded algebra A is written as a quotient $k[x_0, ..., x_n]/I$, where I is an ideal generated by some homogeneous polynomials $f_1, ..., f_r$, then its associated projective scheme $X = \operatorname{Proj} A$ is the locus of common zeros of $f_1, ..., f_r$ in projective space $\mathbb{P}^n$. Conversely, if we are given a projective scheme X, we can recover a graded algebra A as follows: For $n \gg 0$, A_n is the space of all functions on X with pole of order $\leq n$ at infinity. (The relations between A and $\operatorname{Proj} A$ hold only in large degree.)

In order to proceed, we need to rewrite this description in terms of sections of invertible sheaves. Let L denote the invertible sheaf of locally defined functions on X with pole of order ≤ 1 at infinity. Then $L^{\otimes n}$ is the sheaf of local functions with pole of order $\leq n$, so we can identify global functions with pole of order $\leq n$ at infinity with global sections of this sheaf: For $n \gg 0$, $A_n = H^0(X, L^{\otimes n})$. Multiplication in A is induced by the tensor product on L.

Van den Bergh [AV] has shown how to extend this description to construct noncommutative rings. He observes that in order for $L^{\otimes n}$ to be defined, L must have both a left and a right module structure over the structure sheaf $\mathcal{O}_X$, i.e., it must be an $(\mathcal{O}, \mathcal{O})$-bimodule. It is not necessary that the actions on the left and on the right agree; in fact this would be inconsistent if $\mathcal{O}$ weren't commutative. But if L is a bimodule, invertible as left and as right module, then $L^{\otimes n}$ is defined, and setting $B_n = H^0(X, L^{\otimes n})$ yields a graded algebra B which is often noncommutative. Of course, in order that B have reasonable properties, the bimodule L must satisfy a condition analogous to "ampleness" of an invertible sheaf in the commutative case (see [AV]).

The use of bimodules to define a polarization extends to certain noncommutative schemes X as well. But when X is commutative, it is not difficult to see that the left and right actions on an invertible bimodule L differ by an automorphism τ of the scheme X. In other words, the right action will be obtained from the left one by the rule

$$vf = f^\tau v,$$

for $v \in L$ and $f \in \mathcal{O}$. The bimodule obtained in this way from an automorphism will be denoted by L_τ, and we will use the notation $B(X, \tau, L)$ for the algebra defined in this way, i.e., the algebra whose part of degree n is $B_n = H^0(X, L_\tau^{\otimes n})$.

We now consider a noncommutative graded domain A. Thanks to the work of Stafford [ASt], the case that A has GK-dimension 2 is well understood:

Theorem 1.2. *Let B be a graded domain of GK-dimension 2 which is finitely generated by elements of degree 1. Then:*
  (i) *Proj B is a commutative algebraic curve. More precisely, there is a projective algebraic curve C, an automorphism τ of C, and an invertible sheaf L of positive degree on C such that, for large n, $B_n \cong H^0(C, L_\tau^{\otimes n})$.*
  (ii) *The algebra B has the desired properties 1.1.*

Note that this theorem is free of extraneous hypotheses, except for the requirement that A be generated in degree 1, which may seem artificial. In fact, as is explained in [ASt], the situation becomes considerably more complicated when this requirement is dropped.

Following the example of the Italian algebraic geometers at the end of the last century, we may attempt to classify the noncommutative projective surfaces which arise as Proj A, for certain graded domains of GK-dimension 3. The object of this talk is to present a conjecture about them.

2. Examples of graded domains of GK-dimension 3.

There are many examples which show that Theorem 1.2 does not extend directly to higher dimension. It is true that one can construct noncommutative domains of GK-dimension 3 analogous to those of GK-dimension 2 by means of a suitable commutative algebraic surface X, automorphism τ, and invertible sheaf L. But other noncommutative domains exist, and those are the ones that one would like to describe. Here are four basic types:

Example 2.1. *Algebras finite over their centers.*

Algebras which are finite over their centers can be constructed quite simply from orders. Let K be the function field of a projective algebraic surface

S, and let D be a division ring with center K and finite over K. Let $\mathcal{A}$ be an $\mathcal{O}_S$-order in D. One obtains a noncommutative scheme $X = \operatorname{Spec}\mathcal{A}$ by gluing the rings of sections of $\mathcal{A}$ over affine open sets of Z, using central localizations. Then a graded ring A finite over its center can be defined as follows: Let L denote the sheaf $\mathcal{O}(1)$ on S, and set $\mathcal{A}(n) = \mathcal{A} \otimes_{\mathcal{O}} L^{\otimes n} \approx \mathcal{A}(1) \otimes_{\mathcal{A}} \otimes \cdots \otimes_{\mathcal{A}} \mathcal{A}(1)$. Then $A_n = H^0(S, \mathcal{A}(n))$. Other rings, not necessarily finite over their centers, can be constructed using $(\mathcal{A}, \mathcal{A})$-bimodules which which are not central.

Example 2.2. *Auslander regular algebras of dimension* 3.

A graded domain A of finite global dimension which satisfies the Auslander conditions is called an Auslander regular algebra (see[Le]). The Auslander regular algebras of dimension 3 have been classified completely [ASch,ATV], and Schelter's *Sklyanin algebras* are the most interesting ones among them. These are the three-dimensional analogues of some well-known four-dimensional algebras defined by Sklyanin [Skl]. A Sklyanin algebra $A = A(E, \sigma)$ of dimension 3 can be defined in a somewhat mysterious way, in terms of an elliptic curve E embedded as a cubic in $\mathbb{P}^2$ and a translation σ of E. The associated projective scheme $\operatorname{Proj} A$ is a deformation of the projective plane: a *quantum plane* [Ar]. The other Auslander regular algebras of dimension 3, and their associated quantum planes, are obtained from automorphisms of singular plane cubics.

Example 2.3. *Polynomial extensions of domains of GK-dimension* 2.

Here B is a graded domain of GK-dimension 2 and $A = B[z]$, where z is a central variable of degree 1. By Theorem 1.2, B has the form of a twisted homogeneous coordinate ring of a curve: $B = B(C, \tau, L)$. If the automorphism τ of C has infinite order, then neither B nor A is PI. In this case, C will be rational or elliptic, and if it is elliptic, then τ will be a translation. The case that τ is a translation of infinite order on an elliptic curve is especially interesting.

Example 2.4. *Homogenized differential operator rings.*

This example is due to Stafford. Let C be a nonsingular curve with structure sheaf $\mathcal{O}$, and let $\mathcal{D}$ denote the sheaf of rings of differential operators on C. If x denotes a local parameter at a point $p \in C$, then locally at p, $\mathcal{D}$ has the form $\mathcal{O}_p\langle y \rangle$, where y is the derivation $\frac{d}{dx}$, and $yx = xy + 1$. We choose a point p_0 on C, and consider the subsheaf $\mathcal{D}'$ of $\mathcal{D}$ which is equal to $\mathcal{D}$ except at the point p_0, and which is generated by $y_0 = x_0\frac{d}{dx_0}$ at that point, x_0 being a local parameter. Thus the relation $y_0x_0 = x_0y_0 + x_0$ holds at p_0. This relation shows that, locally, x_0 is a normal element of $\mathcal{D}'$, which is the reason that we replace $\mathcal{D}$ by $\mathcal{D}'$. We homogenize the defining relation of $\mathcal{D}'$ using a central variable z, to obtain a sheaf of graded rings $\mathcal{A}$ which has the local form $\mathcal{O}_p\langle y, z\rangle$, with the defining relation $y_0 x_0 = x_0 y_0 + x_0 z$ at p_0 and $yx = xy + z$ at all other points. The subsheaf $\mathcal{A}_n$ of elements of degree n is isomorphic to the sheaf of differential operators which are in $\mathcal{D}'$ and which have order $\leq n$. Locally at the point p_0, the element x_0 normalizes $\mathcal{A}$. We choose an integer $r > 2g - 2$ and we denote by $\mathcal{L}_n$ the right $\mathcal{O}$-module of sections of $\mathcal{A}_n$ with pole of order nr at p_0. So locally at p_0, a section of $\mathcal{L}_n$ can be written in the form $\alpha_n x_0^{-nr}$, where $\alpha_n \in \mathcal{A}_n$. Because x_0 is normalizing, multiplication sends $\mathcal{L}_m \times \mathcal{L}_n \to \mathcal{L}_{m+n}$, so we obtain a graded ring A by setting $A_n = H^0(C, \mathcal{L}_n)$.

There are many other interesting graded domains of dimension 3, for example the *quantum quadrics* which are obtained from Sklyanin algebras of dimension 4 by dividing by a central element of degree 2 [Sm, SStd].

3. A General Description of $X = \operatorname{Proj} A$.

Let A be a noetherian graded ring, and let $\mathcal{C}$ denote the category of finitely generated, graded, right A modules, modulo the subcategory of torsion modules (modules finite-dimensional over k). This category $\mathcal{C} = \operatorname{gr-}A/(torsion)$ can also be described as the category of *tails* $M_{\gg 0}$ of finitely generated graded A-modules. By definition, the projective scheme $X = \operatorname{Proj} A$ associated to A is the triple $(\mathcal{C}, \mathcal{O}, s)$, where $\mathcal{O}$ is image in $\mathcal{C}$ of the right module A_A, and s is the autoequivalence of $\mathcal{C}$ defined by the shift operator on graded modules [AZ,Ma,Ve]. Working out the consequences of this definition is an ongoing program, and we will not need to consider it in detail. However, we need to review some geometric concepts, namely of points and fat points. Following tradition, we assume that $X = \operatorname{Proj} A$ is smooth. This means that $\mathcal{C}$ has finite injective dimension, or that for every finite graded A-module M and for $q > 2$, the graded injectives I^q in a minimal resolution $0 \to M \to I^0 \to I^1 \to \cdots$ are sums of the injective hull of k.

Our conditions 1.1 imply that the ring A can be recovered, in sufficiently large degree, from its associated projective scheme. This means that Zhang's condition χ, that $\operatorname{Ext}^q(k, M)$ is finite dimensional for all finite modules M, holds for A [YZ].

The tail of a critical module M of GK-dimension 1 is called a *fat point* of X. Fat points are the projective analogs of finite dimensional representations of a ring [Sm,SSts]. If the stable Hilbert function $\dim M_n, n \gg 0$, is the constant function 1, so that M has multiplicity 1, then the tail $M_{\gg 0}$ is called a *point* of X.

We say that X is finite over its center if there is a commutative algebraic surface S and a coherent $\mathcal{O}_S$-algebra $\mathcal{A}$ such that X is isomorphic to the relative scheme $\underline{\text{Spec}}\,\mathcal{A}$ over S, as in Example 2.1. Specifically, this means that $\mathcal{C}$ is equivalent to the category of coherent as $\mathcal{O}_S$-modules with a right $\mathcal{A}$-module structure, and that s is an ample autoequivalence of $\mathcal{C}$.

In all known examples of graded domains of GK-dimension 3 which satisfy 1.1 and such that $X = \operatorname{Proj} A$ is *not* finite over its center, the scheme X has some remarkable properties:

3.1. There is a quotient $B = A/I$ of A of pure GK-dimension 2, such that:
  (i) $Y = \operatorname{Proj} B$ is a commutative projective curve, possibly reducible, whose points are points of X.
 (ii) X has only finitely many fat points in addition to the points on Y. In particular, X has only finitely many fat points of multiplicity $\neq 1$.
(iii) The complement of Y in X is an affine open subscheme. In other words, there is a finitely generated noetherian domain R such that $X - Y = \operatorname{Spec} R$, or that the category mod-R of finite R-modules is equivalent with the quotient category gr-$A/(I - \text{torsion})$.

A priori, it is not clear that $\operatorname{Proj} A$ should have any points at all. This is a puzzling point.

4. A Conjecture.

The graded quotient ring of a graded Ore domain A has the form $Q(A) = D[z, z^{-1}, \phi]$, where D is a division ring and ϕ is an automorphism of D [NV]. We will refer to the division ring D as the *function field* of the scheme $X = \operatorname{Proj} A$, and we will say that two such schemes X, X' are *birationally equivalent* if their function fields are isomorphic extensions of k. If X is birational to a quantum plane (Example 2.2) we call it a *q-rational* surface. If X is birational to one of the surfaces listed in Examples 2.3,2.4 and in which the curve C has genus $g > 0$, we call it *q-ruled*. (There is an intrinsic definition of *q*-ruled surface in terms of "bimodule algebras" over a curve [VdB1].)

Conjecture 4.1. *Let k be an algebraically closed field of characteristic zero, and let A be k-algebra of GK-dimension 3 satisfying the properties 1.1. Then $X = \operatorname{Proj} A$ is birationally equivalent to $\operatorname{Proj} A'$, where A' is one of the graded domains described in Examples 2.1-2.4. So one of the following holds:*
  (i)   *X is finite over its center,*
 (ii)   *X is q-rational, or*
(iii)   *X is q-ruled.*

The properties 1.1 are included as hypotheses in this conjecture. Ideally, we would like them to be consequences of more basic assumptions on the structure of A, as is the case in dimension 2 (see Theorem 1.2). However, we don't know what the necessary assumptions are. At this stage of our knowledge, any reasonable hypotheses on the structure are acceptable.

A finer classification would subdivide (i) into two classes:

4.2.
(ia) A finite over its center, and
(ib) A not finite over its center, but X finite over its center.

The possibilities for (ib) can be enumerated conjecturally as well. They include cases in which X is a commutative surface, such as an abelian surface, which has a continuous group of automorphisms. (See the last section of [AV] in this connection.)

It is interesting to note that PI algebras appear as a natural class of rings in Conjecture 4.1. Indeed, the PI case (ia) should be viewed as the "general" one. It corresponds roughly to the class of commutative surfaces of Kodaira dimension ≥ 0, though PI algebras also appear as special cases in (ii),(iii). There is no hope of listing these rings.

For those of us who are interested in describing noncommutative phenomena, it may, at first glance, seem a bit disappointing to think that rational and ruled surfaces could be the only ones which have noncommutative analogues not finite over their centers. One must remember that the most beautiful results of the Italian school, such as the numerical characterizations of rational and ruled surfaces of Castelnuovo and Enriques, and Castelnuovo's theorem on the rationality of plane involutions, concern precisely these surfaces. Whether the conjecture is correct or not, extending those results to the noncommutative setting is a worthy goal for people working in ring theory.

5. The Division Rings.

Since Conjecture 4.1 concerns only the birational equivalence classes of noncommutative surfaces, it can be stated in terms of their function fields. Here is a list of the division rings which are predicted by the conjecture:

List of Division Rings 5.1. *In this list, k is assumed algebraically closed, of characteristic zero, σ denotes a translation by a point of infinite order of an elliptic curve E, and $q \in k^*$ is not a root of unity.*

1. *division rings which are finite algebras over function fields of transcendence degree 2.*

2. *q-rational division rings:*
 (a) *$k_q(x,y)$, the field of fractions of the q-plane $yx = qxy$.*

(b) the Sklyanin division ring $S(E, \sigma)$, the degree zero part of the graded field of fractions of the Sklyanin algebra $A(E, \sigma)$.

(c) D_1, the field of fractions of the Weyl algebra.

3. q-ruled division rings:

(a) $K(E, \sigma)$, the field of fractions of the Ore extension $k(E)[t, \sigma]$.

(b) $D(C)$, the field of fractions of the ring of differential operators on a curve C of positive genus.

It is interesting to note that Schelter's 3-dimensional Sklyanin algebras provide the only division rings on our list which are relatively new. In fact, Van den Bergh showed recently that $S(E, \sigma)$ is the ring of invariants in $K(E, \sigma)$ under the involution defined by the map which sends $p \mapsto -p$ in the group E, and $t \mapsto t^{-1}$. Thus the Sklyanin division rings are also closely related to more classical ones.

There are several definitions of dimension for division rings. The first one, the GK transcendence degree, was introduced by Gelfand and Kirillov [GK,Z1]. They used this notion to distinguish the fields of fractions of the Weyl algebras A_n. One can also define the dimension of D to be its projective dimension as a module over $D \otimes D^{opp}$ [Re,Ro,St]. Recently Zhang [Z2] found an elegant definition (ZD) for which it is easier to prove some general properties: Let D be a division ring over a field k. Then $zd(D) \geq r$ if there exists a finite dimensional k-subspace V of D containing 1 and a constant c such that for every finite dimensional subspace W of D,

$$\dim(VW) \geq \dim(W) + c \dim(W)^{\frac{r-1}{r}}.$$

As Zhang points out, the way to understand this definition intuitively is to imagine $\dim(W)$ as the volume of a variable region in an r-dimensional space. Then with an appropriate constant factor, $\dim(W)^{\frac{r-1}{r}}$ is a lower bound for the volume of the boundary ∂W. The regions VW and W differ only near the boundary.

Zhang has shown that $zd(D)$ is at most equal to the GK transcendence degree. It is not known whether or not the two are always equal. We will call a Z2 division ring one which is finitely generated over k and such that $zd(D)$ is equal to 2.

Conjecture 5.2. *The list 5.1 contains all Z2 division rings.*

Zhang has shown that the division rings listed are distinct, and that certain inclusions among them can not occur. For example, $k_q(x, y)$ is not a subfield of D_1. (Some of these facts were known before.) A convenient tool for verifying them is the concept of *prime divisor*. A prime divisor of a Z2-division ring D is a discrete valuation ring R whose residue field is a function field in one variable over k. The components of the point locus Y of $X =$ Proj A (see 3.1) define valuations of its function field, the ones classically referred to as being "of the first kind" on X.

Proposition 5.3. *Let D be one of the division rings 5.1, and suppose that D is not finite over its center. Then D has at least one prime divisor. More precisely,*

(i) *The prime divisors of $k_q(x, y)$ are determined by the values $v(x), v(y)$, which can be any pair of relatively prime integers. The residue field of every prime divisor is a rational function field.*

(ii) *The Sklyanin division ring $S(E, \sigma)$ has exactly one prime divisor. Its residue field is the function field $k(E)$ of the elliptic curve.*

(iii) *The residue field of every prime divisor of D_1 is a rational function field.*

(iv) *The division ring $K(E, \sigma)$ has exactly two prime divisors. Both have the function field $k(E)$ as residue field.*

(v) *There is exactly one prime divisor of $D(C)$ whose residue field is the function field $k(C)$. All other prime divisors have rational residue fields.*

A prime divisor R comes equipped with an outer automorphism, which is defined by conjugating by a generator of its maximal ideal M, and which provides further information about the division ring. It also has an *index*, the largest integer n such that R/M^n is commutative.

Assertion 5.3(i) is due to Zhang, and Willaert [W] has studied prime divisors in D_1. A proof of 5.3(ii) is outlined in Section 6. We don't know how to prove the existence of a prime divisor directly from the Z2 condition, and indeed, even if one always exists, it may be difficult to give a direct proof because the assertion is false when the assumption that D is not finite over its center is removed. A proof of the existence might be a starting point for classifying the Z2 division rings.

Some heuristic evidence for the conjecture that a Z2 division ring has a prime divisor is provided by the following construction, which will produce one in a few cases: Choose generators for a convenient finitely generated subring S of D, and form a graded ring A by homogenizing the defining relations, using a central variable z of degree 1. If we are very lucky, A will be noetherian and of GK-dimension 3, and z will generate a completely prime ideal P. Then Theorem 1.2 identifies A/P as a twisted homogeneous coordinate ring of a curve. In that case P will be localizable, and the graded localization A_P will be a graded valuation ring, whose subring of degree zero is the required prime divisor.

The set $\mathcal{P}_1$ of prime divisors of the Weyl skew field D_1 forms a fairly complicated picture, but one can give a combinatorial description in terms of the birational geometry of the ordinary projective plane $\mathbb{P}^2$ (see also [W]). Consider prime divisors of the rational function field $k(x, y)$ which are centered on the line L at infinity in $\mathbb{P}^2$. Define the index of such a prime divisor to be the order of pole of the double differential $dx\,dy$, and let $\mathcal{P}_2$ denote the set of prime divisors in $k(x, y)$ of positive index.

Proposition 5.4. *There is a bijective map* $\mathcal{P}_1 \to \mathcal{P}_2$ *which preserves index.*

A similar description can be given for the prime divisors of $D(C)$. This proposition was proved in joint work with Stafford. It is not very difficult, but is too long to include here.

6. Evidence.

Besides the examples, Conjecture 4.1 is based on evidence collected by three methods:

6.1.
(1) quantization, or deformation of commutative schemes,
(2) the theory of orders and the Brauer group, and
(3) Van den Bergh's notion of noncommutative blowing up.

We have no additional evidence on which the rash Conjecture 5.2 that our list of Z2 fields is complete can be based. It is just that no other division rings have appeared up to now.

Discussion of the evidence.

(1) It is reasonable to suppose that a sizable family of noncommutative surfaces would leave a trace as a "classical limit", a commutative scheme. If the conjecture is correct, then the limit surface must be rational or ruled. We may test this conclusion by studying infinitesimal deformations of a commutative surface. As is well known, the main invariant of a first order deformation of a commutative surface X_0 is its Poisson bracket, which is a section of the anticanonical bundle $\wedge^2 T_{X_0} = \mathcal{O}_{X_0}(-K)$. On many surfaces, this bundle has no sections. The first assertion of the following proposition follows directly from the classification of commutative surfaces (see [Be,BPV]). A proof of the second assertion is outlined in Section 8.

Proposition 6.2. *Let* X_0 *be a smooth projective surface which admits a noncommutative infinitesimal deformation* X. *Then*
(i) X_0 *has an effective anticanonical divisor, and is one of the following: a rational surface, a birationally ruled surface, an abelian surface, or a K3 surface.*

(ii) *If there exists an ample invertible sheaf on X_0 which extends to an invertible bimodule on X, then X_0 is rational or birationally ruled.*

The existence of an ample invertible bimodule is necessary in order for the polarization of X_0 to extend to the deformation, i.e., for the homogeneous coordinate ring to deform compatibly (see 8.2 for a precise statement). Thus the classical limit surfaces are of the expected types.

As is well known, the anticanonical divisors on a surface have arithmetic genus 1. Those on ruled surfaces are described by the next proposition.

Proposition 6.3. *Let Z be an effective anticanonical divisor on a ruled surface X over a curve C of genus g.*
  (i) *If $g > 1$, $Z = 2D + F$, where D is a section and $F = \sum F_i$ is a sum of rulings.*
  (ii) *If $g = 1$, then either Z has the above form, or else Z is the sum of two disjoint sections.*

Examples 2.3 and 2.4 are deformations of commutative surfaces determined by Poisson brackets of the forms (ii) and (i) respectively.

(2) Studying orders can provide heuristic evidence for the conjecture that various q-rational surfaces which arise, for example by quantization, should be birationally equivalent. To obtained this evidence, we specialize q to a root of unity or σ to a translation of finite order. Then, in all cases which have been investigated, the algebra A becomes finite over its center, and one can test birational equivalence using known results about the Brauer group. The description of deformations of orders is still being worked out, but Ingalls has shown that if a maximal order whose center is a smooth surface Z admits a non-PI deformation, then the anticanonical sheaf $\mathcal{O}_Z(-K)$ must have a nonzero section which vanishes on the ramification locus of the order. So the center Z of $X = \operatorname{Proj} A$ is one of the surfaces listed in 6.2(i). For instance, Z may be a rational surface with effective anti-canonical divisor, and the anticanonical divisor may be an elliptic curve. The next proposition is rather easy to prove:

Proposition 6.4. *A smooth elliptic curve E has an essentially unique embedding as a cubic curve in $\mathbb{P}^2$. Suppose that E is also embedded as an anticanonical divisor E_1 into a rational surface X_1. Then the pair $E_1 \subset X_1$ is birationally equivalent to the embedding $E_2 \subset X_2$ of E as a plane cubic in $X_2 = \mathbb{P}^2$. In other words, the local rings of X_i at the general points of E_i are isomorphic.*

A similar result holds when the anticanonical divisor E is a cycle of rational curves [Lo]. Now if E is an elliptic curve on a rational surface X and if E'/E is an etale covering of elliptic curves, then Brauer group computations [AM] show that there is a division ring D with center the rational function field $k(X)$, whose branching data is this given covering, and that D is unique up to $k(X)$-isomorphism. The proposition shows that, provided that E is anticanonical, D is isomorphic to the division ring obtained from the cubic embedding $E \subset \mathbb{P}^2$. This is what the conjecture would predict for the Sklyanin division ring, if σ were allowed to have finite order.

(3) Having plausibility arguments for the existence of birational maps between certain of the projective schemes, a natural question is: What are these birational maps? In the commutative case, a theorem of Zariski asserts that one can factor any birational transformation between smooth surfaces into a succession of blowings up and down. The key ingredient which has been provided by Van den Bergh [VdB2] is to describe the noncommutative analogue of the blowing up of a surface. He has shown that in favorable cases one can blow up a point p of the point locus of X, obtaining another projective scheme X' in which the point p is replaced by an exceptional module. He has also shown how the blowing up process produces the mysterious sporadic fat points which appear on special quantum quadrics (see [S],[SSts]).

Because blowing up is an essentially projective construction, the definition is subtle and the blowing up does not lead to a projective scheme in all cases. We refer to Van den Bergh's paper for the definitions. For the purposes of this paper, it seems sufficient to illustrate the process by an example. This is done in Section 9.

7. Prime Divisors of the Sklyanin Division Ring:

This section gives a proof of Proposition 5.3 (ii). We refer to the literature for known results about the 3-dimensional Sklyanin algebra $A(E, \sigma)$. Recall that σ is assumed of infinite order. Let A denote the 3-Veronese of $A(E, \sigma)$. There is a central element g of $A(E, \sigma)$ of degree 3 [ATV], and it has degree 1 in A. Let $Q = D[z, z^{-1}; \tau]$ denote the graded fraction field of A. One can take for z any element of A_1 (or of Q_1 for that matter). A change of the element $z \in Q_1$ changes τ by an inner automorphism. Since g is central, $\tau = 1$ when $z = g$. Thus τ is inner for all choices of z.

Let R be a prime divisor of D, a discrete valuation whose residue field $K = R/M$ is a function field in one variable, and let ν denote the associated valuation. A discrete valuation is stable under inner automorphism. Thus R is τ-stable, and $R[z, z^{-1}; \tau]$ is defined and is a subring of Q.

Lemma 7.1. *We may choose* $z \in A_1$ *so that* $A_1 \subset Rz$. *When this is done,* $A \subset R[z; \tau]$.

Proof. With g as above, we choose $u \in A_1 g^{-1}$ with $\nu(u)$ minimal. Then $A_1 g^{-1} u^{-1} \subset R$. So $z = ug$ has the required property.

We denote the automorphism of the residue field K of R which is induced by the action of τ on R by τ too, so that $K[z, z^{-1}; \tau] = R[z, z^{-1}; \tau] \otimes_R K$. Let $\overline{A}$ denote the image of A in $K[z, z^{-1}; \tau]$, and let π be the canonical homomorphism $A \to \overline{A}$. Thus $\overline{A}$ is a graded domain.

Lemma 7.2. $\overline{A} = A/gA$.

Proof. We use the fact that the Sklyanin algebra has no two-sided graded ideal $\overline{I}$ such that $gk(\overline{A}/\overline{I}) = 1$. Since $K[z, z^{-1}; \tau]$ is a domain of GK-dimension 2, $gk(\overline{A}) \leq 2$. It is at least 1 because $z \in A_1$, and it can't be 1. So $gk(\overline{A}) = 2$, and $\overline{A}$ is the coordinate ring of a twisted curve, one of the rings described in Theorem 1.2. We also know that $\overline{A}$ has no graded ideal I such that $gk(A/I) = 1$, because A has none. Therefore every nonzero ideal of $\overline{A}$ is cofinite. Let $\overline{g}$ be the residue in $\overline{A}$ of the central element g. If $\overline{g}$ were not 0, $\overline{A}/\overline{g}\overline{A}$ would have GK-dimension equal to $gk(\overline{A}) - 1 = 1$. Since this is impossible, $\overline{g} = 0$ and g is in the kernel of π. Since g generates a completely prime ideal in A and $gk(A/gA) = 2$, $\overline{A} = A/gA$.

The set S of homogeneous elements of A which are not divisible by g is an Ore set, and the degree zero part of the ring of fractions $S^{-1}A$ is the valuation ring of the g-*adic* valuation which, on A, is given by the rule $\nu(a) = r$ if $a = g^r b$ and g does not divide b. By what has been proved above, $S^{-1}A \subset R[z, z^{-1}; \tau]$, because the image of $s \in S$ in the graded division ring $K[z, z^{-1}; \tau]$ is not zero. Since $(S^{-1}A)_0$ is a valuation ring, this ring must be R.

8. Deformation of a Commutative Surface.

In order to keep the discussion brief, we restrict our attention to first order deformations, those parametrized by the ring $R = k[\epsilon]$, $\epsilon^2 = 0$. We denote by $\mathcal{A}_R$ the category of R-algebras A such that $A \otimes_R k$ is commutative.

Let us call a *scheme* X_R in $\mathcal{A}_R$ a commutative scheme X_0, together with an extension of its structure sheaf $\mathcal{O}_{X_0}$ to a sheaf of rings $\mathcal{O}_X$ in $\mathcal{A}_R$, compatibly with localization. The sheaf $\mathcal{O}_X$ will be called the *structure sheaf* of X. By *coherent sheaf* on an R-scheme X_R, we mean a sheaf of finite right $\mathcal{O}_X$-modules which is compatible with localization. The scheme X is *smooth* if $\mathcal{O}_X$ is flat over R and if X_0 is smooth. We write $\mathcal{O} = \mathcal{O}_X$, $\mathcal{O}_0 = \mathcal{O}_{X_0}$, and we denote the tangent sheaf on X_0 by T_0.

Because X_0 is commutative, the commutator $[x, y]$ on $\mathcal{O}$ can be viewed as a skew symmetric map $\alpha : \mathcal{O}_0 \times \mathcal{O}_0 \to \mathcal{O}_0$ which is a derivation in each variable. We call such a skew derivation a *bracket*. The next proposition is standard.

Proposition 8.1.

(i) Let X_0 be a smooth scheme over k. The set of brackets on X_0 is classified by $H^0(X_0, \wedge^2 T_0)$.

(ii) If a bracket α is given, then a smooth extension of X_0 to R with commutator α exists locally. The obstruction to its existence globally lies in $H^2(X_0, T_0)$. If the obstruction vanishes, then the isomorphism classes of of extensions X whose commutators are the given bracket form a principal homogeneous space under $H^1(X_0, T_0)$.

(iii) For any smooth extension X, the sheaf $\underline{\mathrm{Aut}}(\mathcal{O})$ of local automorphisms of X which reduce to the identity on X_0 is isomorphic to T_0.

The first assertion of Proposition 6.2 follows from (ii) and the classification of surfaces [Be,BPV].

Suppose that a smooth extension X is given, and that X_0 is projective, with ample line bundle L_0. We consider the problem of extending this polarization to X, so as to obtain a noncommutative projective scheme in the sense of [AZ]. What we want is an R-linear, ample autoequivalence s of the category mod-X of coherent sheaves over X which extends the polarization $s_0 = \cdot \otimes_{\mathcal{O}_0} L_0$ of X_0 defined by L_0. We call an $(\mathcal{O}, \mathcal{O})$-bimodule L *invertible* if R acts centrally on L, L is locally isomorphic to $\mathcal{O}$ as left and as right module, and $L_0 = L \otimes_R k$ is a central $\mathcal{O}_0$-bimodule.

Proposition 8.2. Let X be a scheme in $\mathcal{A}_R$, and let L_0 be an invertible sheaf on X_0. Let s be an autoequivalence of mod-X which extends the autoequivalence s_0 of mod-X_0 defined by L_0.

(i) There is an invertible $\mathcal{O}$-bimodule L such that $s \cong \cdot \otimes_{\mathcal{O}} L$ and $L \otimes_R k \approx L_0$. If s_0 is ample, then so is s.

(ii) With L as above, set $A = \bigoplus H^0(X, L^{\otimes n})$ and $A^0 = \bigoplus H^0(X_0, L_0^{\otimes n})$. Then A is a noetherian graded R-algebra, $A_{\gg 0}$ is R-flat, and $A_{\gg 0} \otimes k \approx A_{\gg 0}^0$.

We analyze the problem of extending $\mathcal{O}_0$-bimodule L_0 to $\mathcal{O}$ in two steps. First, we extend the right module structure. Right $\mathcal{O}$-modules locally isomorphic to $\mathcal{O}$ are classified by $H^1(X, \mathcal{O}^*)$, and there is an exact sequence

$$0 \to \mathcal{O}_0^+ \xrightarrow{1+\epsilon} \mathcal{O}^* \to \mathcal{O}_0^* \to 0.$$

Thus, as in the commutative case, the obstruction to extending the right module L_0 lies in $H^2(X, \mathcal{O}_0)$, and if it is zero, then the group $H^1(X, \mathcal{O}_0)$ operates transitively on the set of classes of extensions. This is a standard situation.

Next, we consider the left module structure of an invertible right module $L_{\mathcal{O}}$. The commutant $\mathcal{E} = \underline{\mathrm{End}} L_{\mathcal{O}}$ is locally isomorphic to $\mathcal{O}$. More precisely, if $\lambda \in H^1(X, \mathcal{O}^*)$ is the class representing $L_\mathcal{O}$, then $\mathcal{E}$ is the associated "inner form" of $\mathcal{O}$, defined by conjugation by λ. It is not hard to see that if $L_\mathcal{O}$ has the structure of an invertible bimodule which is compatible with the central bimodule structure on $\mathcal{L}_0$, then the commutant $\mathcal{E}$ must be isomorphic to $\mathcal{O}$ by an isomorphism which reduces to the canonical isomorphism $\mathcal{E}_0 \cong \mathcal{O}_0$, and this means that the image of λ in $H^1(X, \underline{\mathrm{Aut}}\,\mathcal{O}) = H^1(X_0, T_0)$ must be the trivial class. This is difficult to achieve when the sheaf $\mathcal{O}$ is noncommutative.

Let the class associated to λ in $H^1(X, \underline{\mathrm{Aut}}\,\mathcal{O})$ be denoted by c_λ. To analyze this image, we suppose that the ground field k is the field $\mathbb{C}$ of complex numbers. Then we may pass to the category of analytic sheaves, and we do so, retaining the same notation. When written in terms of a global section α of $\wedge^2 T_0$, the commutation law in $\mathcal{O}$ becomes the following: Let u, v be the residues in $\mathcal{O}_0$ of elements $u_1, v_1 \in \mathcal{O}$. Then $[u_1, v_1] = \langle \alpha, du \wedge dv \rangle \epsilon$, where $\langle \cdot, \cdot \rangle$ denotes the pairing $\wedge^2 T_0 \times \Omega_0^2 \to \mathcal{O}_0$. The section α also defines a map $\widetilde{\alpha} : \Omega_0^1 \to \Omega_0^{1\vee} = T_0$, by

$$(8.3) \qquad \widetilde{\alpha}(\eta) = \langle \alpha, \eta \wedge \cdot \rangle.$$

Direct computation shows that there is a row-exact diagram of analytic sheaves

$$(8.4)$$

$$
\begin{array}{ccccccccc}
0 & \longrightarrow & \mathbb{C}^* & \longrightarrow & \mathcal{O}_0^* & \xrightarrow{\;dlog\;} & \Omega_0^1 & \xrightarrow{\;d\;} & \Omega_0^2 & \longrightarrow & 0 \\
& & \| & & \| & & \widetilde{\alpha}\downarrow & & \downarrow & & \\
0 & \longrightarrow & \mathbb{C}^* & \longrightarrow & \mathcal{O}_0^* & \xrightarrow{\;c\;} & T_0 & \longrightarrow & \underline{\mathrm{Out}}(\mathcal{O};\mathcal{O}_0) & \longrightarrow & 0
\end{array}
$$

where the top line is the logarithmic de Rham complex, and $\underline{\mathrm{Out}}(\mathcal{O}, \mathcal{O}_0)$ denotes the sheaf of outer automorphisms which reduce to the identity on $\mathcal{O}_0$. This diagram allows us to interpret the obstruction c_λ in terms of the pairing $\wedge^2 T_0 \times \Omega_0^1 \to T_0$ defined by 8.3. We have

$$(8.5) \qquad c_\lambda = \widetilde{\alpha}\, dlog\lambda_0 = \alpha \cup dlog\lambda_0.$$

Note also that the map $H^1(dlog)$ factors through the Chern class map:

$$H^1(X_0, \mathcal{O}^*) \to H^2(X_0, \mathbb{C}) \to H^1(X_0, \Omega_0^1).$$

The image of the Chern class map is contained in $H^{1,1}$, which implies that the second of these maps is injective on the image. Hence $dlog(\lambda_0)$ is not zero unless its Chern class vanishes, which is not the case if λ_0 is the class of an ample invertible sheaf.

Proof of Proposition 6.2(ii). Because of Proposition 6.2(i), we need only rule out the possibility that X is a minimal model of a K3 surface or an abelian surface. In these cases, $\wedge^2 T_0 \cong \mathcal{O}_0$. The nonzero global section $\alpha \in H^0(X_0, \wedge^2 T_0)$ is unique up to scalar factor, and is nowhere zero. Therefore the map $\tilde{\alpha}$ is bijective, and the rows of 8.4 are isomorphic. If λ_0 is the class of an ample invertible sheaf L_0, then $dlog(\lambda_0) \neq 0$, hence $c_\lambda \neq 0$.

Note that the tensor product of two invertible bimodules on X is also invertible, as is the dual module $L^\vee = \operatorname{Hom}(L, \mathcal{O})$. Hence the subset of $\operatorname{Pic} X_0$ which consists of the classes of invertible sheaves on X_0 which admit an invertible extension to X is a subgroup.

Proposition 8.6. *Suppose that $X \in \mathcal{A}_R$ is smooth, of dimension d. Then the class of the canonical sheaf Ω_0^d extends to an invertible bimodule on X.*

A *del Pezzo* surface is a surface X_0 on which $\wedge^2 T_0$ is ample.

Proposition 8.7. *A del Pezzo surface X_0 admits an extension to R which is not commutative, and every such extension has an ample invertible bimodule.*

Proof. Such a surface is rational, and Riemann-Roch computations show that $H^2(X_0, T_0) = 0$, $H^q(X_0, \wedge^2 T_0) = 0$ if $q \neq 0$, while $H^0(X_0, \wedge^2 T_0) \neq 0$. Thus the obstructions 8.1 to extending a bracket α and a ring $\mathcal{O}$ to R vanish.

9. An example of noncommutative blowing up.

We have chosen an example in which we blow up a point of an affine surface $X = \operatorname{Spec} R$. The blown up surface will have the form $X' = \operatorname{Proj} B$, where B is a graded R-algebra. We take $R = k\langle x, y \rangle$, with commutation relation $yx = (x + 1)y$. The point to be blown up is the one defined by the two-sided ideal $\mathfrak{m} = (y, x)R$. Note that the Rees ring $R \oplus \mathfrak{m} \oplus \mathfrak{m}^2 \oplus \cdots$, which defines the blowing up in the commutative case, will not work here because, as is easily seen, $y \in \mathfrak{m}^n$ for all n. Van den Bergh proceeds by twisting this construction.

Let σ denote the automorphism of R of conjugation by y^{-1}, i.e., $y^\sigma = y$, and $x^\sigma = x - 1$. Let $R\langle t; \sigma \rangle$ denote the skew polynomial ring with commutation relation $tr = r^\sigma t$, and extend σ to $R\langle t; \sigma \rangle$ by setting $t^\sigma = t$. Let $\mathfrak{m}_i$ denote the two-sided ideal $(y, x - i)R$. Then $t\mathfrak{m}_i = \mathfrak{m}_{i+1}t$ and $\mathfrak{m}_i^\sigma = \mathfrak{m}_{i+1}$.

The blowing up of $X = \operatorname{Spec} R$ is $X' = \operatorname{Proj} B$, with $B = R\langle v, w \rangle \subset R\langle t; \sigma \rangle$, where $v = tx$ and $w = ty$. Then

$$(9.1) \qquad B = R \oplus \mathfrak{m}_1 t \oplus \mathfrak{m}_1 \mathfrak{m}_2 t^2 \oplus \mathfrak{m}_1 \mathfrak{m}_2 \mathfrak{m}_3 t^3 \oplus \cdots$$

The set

$$(9.2) \qquad \{y^n, (x-1)y^{n-1}, (x-1)(x-2)y^{n-2}, ..., (x-1)\cdots(x-n)\}$$

generates $\mathfrak{m}_1 \cdots \mathfrak{m}_n$, and its residues form a basis of $(\mathfrak{m}_1 \cdots \mathfrak{m}_n)/(\mathfrak{m}_0 \cdots \mathfrak{m}_n)$. The element w is central in B.

The role of the exceptional curve, the inverse image of the point in the commutative case, is played here by the "exceptional" right module $E_B = B/\mathfrak{m}_0 B$. One obtains a basis of E_n by multiplying the residues of (9.2) by t^n. So the Hilbert function of E is $dim_k E_n = n + 1$. The exceptional module E has a unique point module $P = E/wE$ as quotient. The residue of the element $(x-1)\cdots(x-n)$ is a basis of P_n.

The scheme X' has one affine open set, namely $\mathrm{Spec}\, R'$, where $R' = B[w^{-1}]_0$. Setting $z = w^{-1}v = y^{-1}x$, one finds $R' = k\langle y, z\rangle/(zy = yz - 1)$: the Weyl algebra. The restriction of the exceptional module to this affine is the module R'/yR', which is the standard simple module over R'.

References.

[Aj] K. Ajitabh, Residue complex for Sklyanin algebras of dimension three, Adv. in Math. (to appear)

[ASZ] K. Ajitabh, S. P. Smith, J. J. Zhang, Auslander-Gorenstein rings and their injective resolutions (preprint).

[Ar] M. Artin, Geometry of quantum planes, Contemporary Math. 124, 1992 1-15.

[AM] M. Artin and D. Mumford, Some elementary examples of unirational varieties which are not rational, Proc. London Math. Soc. 3 (1972), 75-95.

[ASch] M. Artin and W. Schelter, Graded algebras of global dimension 3, Advances in Math. 66 (1987), 171-216.

[ASt] M. Artin and J. T. Stafford, Noncommutative graded domains with quadratic growth, Inventiones Math. 122 (1995), 231-276.

[ATV] M. Artin, J. Tate and M. Van den Bergh, Some algebras associated to automorphisms of elliptic curves, The Grothendieck Festschrift, Birkhauser, Basel, 1990.

[AV] M. Artin and M. Van den Bergh, Twisted homogeneous coordinate rings, J. Algebra 133 (1990), 249-271.

[AZ] M. Artin and J. J. Zhang, Noncommutative projective schemes, Adv. in Math. 109 (1994), 228-287.

[BPV] W. Barth, C. Peters, and A. Van de Ven, Compact Complex Surfaces, Springer-Verlag, Berlin 1984.

[Be] A. Beauville, Surfaces algébriques complexes, Astérisque 54, Soc. Math. France 1978.

[Bj] J.-E. Björk, The Auslander condition on noetherian rings, Séminaire Dubreil-Malliavin 1987–88, Lecture Notes in Math. 1404, Springer Verlag (1989) 137–173.

[GK] I. M. Gelfand and A. A. Kirillov, Sur les corps liés aux algèbres enveloppantes des algèbres de Lie, Pub. Math. Inst. Hautes Etudes Sci. 31 (1966), 5-19.

[Le] T. Levasseur, Some properties of non-commutative regular rings, Glasgow J. Math. 34 (1992) 277-300.

[Lo] E. Looijenga, Rational surfaces with an anti-canonical cycle, Annals of Math. 114 (1981),267-322.

[Ma] Yu. I. Manin, Quantum Groups and Non-Commutative Geometry, Pub. C.R. Math. Univ. Montréal, Montréal 1988.

[MR] J. C. McConnell and J. C. Robson, Noncommutative Noetherian Rings, Wiley-Interscience, Chichester, 1987.

[NV] C. Năstăsescu and F. Van Oystaeyen, Graded Ring Theory, North Holland, Amsterdam, 1982.

[Re] R. D. Resco, Transcendental division algebras and simple noetherian rings, Israel J. Math., 32 (1979), 235-256.

[Ro] A. Rosenberg, Homological dimension and transcendence degree, Comm. Alg., 10 (1982), 329-338.

[Skl] E. K. Sklyanin, Some algebraic structures connected to the Yang-Baxter equation, Funct. Anal. Appl. 16 (1982), 27-34.

[Sm] S. P. Smith, The four-dimensional Sklyanin algebra, K-theory 8 (1994) 65-80.

[SStd] S. P. Smith and J. T. Stafford, Regularity of the four dimensional Sklyanin algebras, Compositio Math. 83 (1992), 259-289.

[SSts] S. P. Smith and J. Staniszkis, Irreducible representations of the four-dimensional Sklyanin algebra at points of infinite order, J. Algebra 160 (1993) 57-86.

[St] J. T. Stafford, Dimensions of division rings, Israel J. Math, 45 (1983), 33-40.

[VdB1] M. Van den Bergh, A translation principle for the four-dimensional Sklyanin algebras (preprint).

[VdB2] M. Van den Bergh, Noncommutative blowing up (in preparation).

[Ve] A. B. Verevkin, On a noncommutative analogue of the category of coherent sheaves on a projective variety, Amer. Math. Soc. Transl. (2), 151 (1992), 41-53.

[W] L. Willaert, Discrete valuations on Weyl skew fields (preprint).

[Y1] A. Yekutieli, Dualizing complexes over noncommutative graded algebras, J. Algebra 153 (1992), 41–84.

[Y2] A. Yekutieli, The residue complex of a noncommutative graded algebra (preprint).

[YZ] A. Yekutieli and J. J. Zhang, Serre duality for noncommutative projective schemes, Proc. Amer. Math. Soc. (to appear).

[Z1] J. J. Zhang, On Gelfand-Kirillov transcendence degree, Transactions Amer. Math. Soc. (to appear).

[Z2] J. J. Zhang, On lower transcendence degree of noncommutative division algebras, (in preparation).

DEPARTMENT OF MATHEMATICS, MIT, CAMBRIDGE, MA 02139.

The mathematical influence of Maurice Auslander in Mexico.

Raymundo Bautista

Instituto de Matemáticas, U.N.A.M.

México 04510, D.F., México

E-mail adress: raymundo gauss.matem.unam.mx

The first visit of Maurice Auslander to Mexico was in the summer of 1975. He lectured on several subjects in the representation theory of algebras. We were impressed mainly by the part of the lectures related to almost split sequences, then recently discovered by M. Auslander and I. Reiten.

At that time we were interested in the Coxeter and reflection functors introduced by Bernstein-Gelfand-Ponomarev [10].

Apparently there were some connections between Coxeter functors and Dtr. Later on [11] these connections were in fact established.

During $1976 - 1977$ Roberto Matínez and the author spent two years at Brandeis University. There, we had the opportunity of knowing, and living in, an exciting atmosphere. We met many people through Maurice who were interested in the representation theory of algebras.

In the following we recall some of the mathematical results in representation theory obtained in Mexico due to the influence of Maurice Auslander.

1. Almost split sequences and irreducible maps.

Take Λ an artin algebra, denote by $mod\Lambda$ the full subcategory of the category of left $\Lambda-$modules whose objects are finitely generated modules.

We recall the definition of almost split sequence.

1.1 Definition: An exact sequence in $mod\Lambda$:

$$(1) \qquad 0 \longrightarrow A \xrightarrow{i} B \xrightarrow{j} C \longrightarrow 0$$

is said to be almost split sequence if:

i) the sequence does not split,

ii) A and C are indecomposables,

iii) if $f : X \to C$ is nonsplittable mono, there is some $g : X \to B$ such that $jg = f$.

1.2 Theorem (Auslander-Reiten)[2]: *For any nonprojective indecomposable module C in $mod\Lambda$ there is an almost split sequence (1). For any noninjective indecomposable module A in $mod\Lambda$ there is an almost split sequence (1). Moreover there are dualities $D : mod\Lambda \to mod\Lambda^{op}$ and $tr : \underline{mod}\Lambda \to \underline{mod}\Lambda^{op}$ ($\underline{mod}\Lambda$ has the same objects as $mod\Lambda$ and the morphisms are module those factorizing by projectives) such that in (1) $A \cong DtrC$ and $C \cong trDA$.*

A concept related with almost split sequences is the concept of irreducible map.

1.3 Definition: If X and Y are objects in $mod\Lambda$, a map $f : X \to Y$ is called irreducible if:

a) f is neither epi splittable nor mono splittable,

b) if $f = vu$ where $u : X \to Z$ and $v : Z \to Y$ then either u is mono splittable or v is epi splittable.

On can see easily that any irreducible map is either mono or epi. On the other hand in (1) i and j are irreducible maps. Moreover if $g : X \to C$ is an irreducible map, this map is a direct summand of j. Dually if $h : A \to Y$ is an irreducible map then h is a direct summand of i.

If in (1) we decompose B as a sum of indecomposable modules, $B = \coprod_{i=1}^{r} n_i E_i$ with $E_i \not\cong E_j$, we obtain the invariants n_i, r, $\sum n_i$. The simplest question one can ask is the following: for which kind of algebras is there some almost split sequence (1) with B indecomposable?.

M. Auslander and I. Reiten proved that if Λ is of finite representation type, then there is always some almost split sequence with indecomposable middle term [3]; Roberto Martínez proved that this is so for any artin algebra, [17].

In [5] we obtained an interpretation of the numbers n_i in terms of the radical of a category.

We recall that

$$rad(X,Y) = \{f \in Hom(X,Y) \mid 1 - gf \text{ is invertible } \forall g \in Hom(Y,X)\}$$
$$= \{f \in Hom(X,Y) \mid 1 - fh \text{ is invertible } \forall h \in Hom(X,Y)\}$$

Then

$$n_i = dim_{End(E_i)/rad\ End(E_i)}\ rad/rad^2(E_i, C)$$
$$= dim_{End(E_i)/rad\ End(E_i)}\ rad/rad^2(A, E_i)$$

In the Maurice's lectures in Mexico of 1975 he gave the following intringuing result.

1.4 Proposition: *Let* $0 \longrightarrow A \xrightarrow{f} B \xrightarrow{g} C \longrightarrow 0$ *be a non split exact sequence in an arbitrary abelian category* $\mathcal{C}$. *Then:*

a) $f : A \to B$ *is irreducible if and only if* $g : B \to C$ *has the property that given any morphism* $h : X \to C$ *there is either a morphism* $t : X \to B$ *such that* $gt = h$ *or a morphism* $s : B \to X$ *such that* $hs = g$,

b) $g : B \to C$ *is irreducible if and only if the morphism* $f : A \to B$ *has the property that given any morphism* $h : A \to Y$ *there is a morphism* $t : B \to Y$ *such that* $tf = h$ *or a morphism* $s : Y \to B$ *such that* $sh = f$.

Using this proposition and the above description of the numbers n_i one can obtain some restriction for the numbers n_i in the case when Λ is of strongly bounded representation type. We recall the definition: Λ is of strongly bounded representation type if for any given number n there is only a finite number of indecomposable modules having composition series over the center of Λ of length n.

1.5 Proposition: *Suppose* Λ *is an infinite algebra of strongly bounded representation type, then:*

i) if $f : X \to Y$ *is an irreducible map, then either* X *or* Y *is indecomposable,*

ii) if $B = \coprod n_i E_i$ *is the middle term of an almost split sequence with* $E_i \not\cong E_j$ *then* $n_i \leq 3$,

iii) in the above if for some i, $n_i \geq 2$ *then* $n_j = 1$ *for* $i \neq j$,

iv) if Λ *is a finite dimensional algebra over an algebraically closed field then* $n_i = 1$ *for all* i.

Later on this result was improved for the case of finite representation type.

1.6 Theorem (Bautista-Brenner) [6]: *If* Λ *is of finite representation type then* $\sum n_i \leq 4$, *if the equality holds one of the modules* E_i *is projective injective.*

2. Powers of Dtr.

One of the results from the lectures of Maurice Auslander in Mexico which deeply impressed us was one obtained by Maurice and the then student at Brandeis, Maria Inés Platzeck.

2.1 Theorem (Auslander-Platzeck) [1]: *Take an artin hereditary algebra* Λ, *then* Λ *is of finite representation type if and only if for all indecomposable module* M *there is some* n *such that* $Dtr^n M = projective$.

In fact this result was previously proved by Dlab-Ringel [13] using the methods developped by Bernstein-Gelfand-Poromarev in [10]. However the importance of the Auslander-Platzeck result lies in the method of proof, which allowed the study of more general cases.

In Brandeis Maria Inés, Roberto and the author had many and fruitful discussions (in spanish) on these and related topics. Maurice was always very interested in the outcome of such discussions. One of the main properties of hereditary algebras used for proving 2.1 was the fact that maps between indecomposable projective modules are always monomorphisms. So if we consider artin algebras with this property we get a generalization of hereditary algebras: locally hereditary algebras.

2.2 Definition: An artin algebra is called locally hereditary algebra if any local submodule of a projective is projective. Here a local module is one with only one maximal submodule.

2.3 Theorem [6]: *Take Λ a locally hereditary algebra then Λ is of finite representation type if and only if for any indecomposable module M in $mod\Lambda$ there is some n with $Dtr^n M = projective$.*

This kind of algebra includes the algebras of finite representation type considered and classified by M. Loupias [16].

Back in Mexico, Roberto Martínez and the author became interested in the work of the Kiev school of representation theory.

Nazarova and Roiter introduced the concept of representations of a poset. Later on P. Gabriel [14] gave a new definition which we use here.

Let S be a finite partially ordered set and k a field.

2.4 Definition: A representation of S over k is given by a k-vector space V and a function $\varphi : S \rightarrow k-$vector subspaces of V such that if $s_1 \leq s_2$, $\varphi(s_1) \subseteq \varphi(s_2)$. If (V,φ) and (W,ψ) are representations of S, a map $\rho : (V,\varphi) \rightarrow (W,\psi)$ is given by a linear map $\rho : V \rightarrow W$ such that $\rho\varphi(i) \subseteq \psi(i)$ for all $i \in S$. We denote by $Rep(S,k)$ the category of representations of S over k.

The category $Rep(S,k)$ has direct sums. The Krull-Schmidt theorem holds true in $Rep(S,k)$. We will say that S is of finite representation type over k if $Rep(S,k)$ has only a finite number of isomorphism classes of indecomposable objects.

We will say that the sequence

$$(1) \quad 0 \longrightarrow (V_1,\varphi_1) \xrightarrow{\rho_1} (V_2,\varphi_2) \xrightarrow{\rho_2} (V_3,\varphi_3) \longrightarrow 0$$

is an exact sequence in $Rep(S,k)$ if $0 \longrightarrow V_1 \xrightarrow{\rho_1} V_2 \xrightarrow{\rho_2} V_3 \longrightarrow 0$ is exact and for all $s \in S$ the sequence

$$0 \longrightarrow \varphi_1(s) \xrightarrow{\rho_1} \varphi_2(s) \xrightarrow{\rho_2} \varphi_3(s) \longrightarrow 0$$

are exact.

We have in $Rep(S, k)$ a definition of almost split sequences similar to the one in algebras. Moreover if (1) is an almost split sequence there is an operator similar to Dtr, F such that $(V_1, \varphi_1) \cong F(V_3, \varphi_3)$. We have the following:

2.5 Proposition (Bautista-Martínez)[9]: *$Rep(S, k)$ has almost split sequences. Moreover S is of finite representation type if and only if for any indecomposable object $M \in Rep(S, k)$ there is some n with $F^n M = $ projective object in $Rep(S, k)$.*

For the proof we construct $\Lambda(S)$ a $1-$Gorenstein locally hereditary algebra such that $Rep(S, k) \cong ts\Lambda(S)_{\Lambda_S} = $ full subcategory of $mod\Lambda(S)$ with as objects the submodules of projectives which do not contain $\Lambda(S)$ as a direct summand. Following some suggestions of Maurice, we proved that $ts\Lambda(S)_{\Lambda_S}$ has almost split sequences.

3. Almost split sequences in subcategories.

In the summer of 1978 Maurice made his second visit to Mexico. He was very interested in the results surounding prop 2.5. We had several discussions on this point. Our main interest was in the possibility of using the machinary of almost split sequences for the theory of representations of partially ordered sets. Maurice's point of view was more general, to see under what conditions a full subcategory of $mod\Lambda$ has almost split sequences. During his stay in Mexico he presented us with his joint work with S. Smalo on preprojective partitions. Later on they produced their nice work on almost split sequences in subcategories of $mod\Lambda$.

3.1 Definition: Let $\mathcal{C}$ be a subcategory of $mod\Lambda$. Then a morphism $g : B \to C$ in $\mathcal{C}$ is said to be right almost split morphism in $\mathcal{C}$ if:
   i) g is not a splittable epimorphism,
   ii) if $h : C' \to C$ is a non splittable epimorphism, then there is $h' : C' \to B$ with $h = gh'$.

We have the dual definition for $g : C \to B$ left almost split morphism in $\mathcal{C}$.

An object C in $\mathcal{C}$ is called $Ext-$projective in $\mathcal{C}$ if there are not non trivial short exact sequence in $\mathcal{C}$ ending in C :

$$0 \longrightarrow D \longrightarrow E \longrightarrow C \longrightarrow 0$$

A similar definition for Ext−injectives.

Then we say that $\mathcal{C}$ has almost split sequences if:
i) has right and left almost split maps,
   ii) for any C in $\mathcal{C}$ no Ext−projective, there is an almost split sequence
$0 \longrightarrow C' \longrightarrow E \longrightarrow C \longrightarrow 0$,
   iii) for any C in $\mathcal{C}$ no Ext−injective there exists an almost split sequence
$0 \longrightarrow C \longrightarrow E \longrightarrow C' \longrightarrow 0$.

We have the following theorem.

3.2 Theorem (Auslander-Smalo) [4]: *If $\mathcal{C}$ is closed under extensions and functorially finite, then $\mathcal{C}$ has almost split sequences.*

We recall that $\mathcal{C}$ is called covariantly finite if for any X in $mod\Lambda$ there exists W in $\mathcal{C}$ and an epi

$$\eta : Hom(W, -) \mid_{\mathcal{C}} \longrightarrow Hom(X, -) \mid_{\mathcal{C}} \longrightarrow 0$$

The category $\mathcal{C}$ is called contravariantly finite if for any X in $mod\Lambda$ exist Z in $\mathcal{C}$ and an epi

$$\rho : Hom(-, Z) \mid_{\mathcal{C}} \longrightarrow Hom(-, X) \mid_{\mathcal{C}} \longrightarrow 0$$

Finally $\mathcal{C}$ is called functorially finite if is both contravariant and covariantly finite.

With this result in mind, M. Kleiner and the author became interested in the existence of almost split sequences for the representation of bocses. So we joined forces and we proved this existence in 1988.

We recall that a bocs is a coalgebra $\mathcal{A} = (C, \mu, \epsilon)$ over an algebra A (algebra over an algebraically closed field k). Here $\mu : C \rightarrow C \otimes_A C$ is the comultiplication and $\epsilon : C \rightarrow A$ the counit, with the usual properties and some further restrictions.

The category $\mathcal{R}$ of representations of $\mathcal{A}$ has as objects the left A−modules and maps $\mathcal{R}(M, N) = Hom_A(C \otimes_k M, N)$.

If $C \otimes M \xrightarrow{g} N$ and $h : C \otimes N \xrightarrow{f} L$ are maps in $\mathcal{R}$ its composition is given by

$$C \otimes M \xrightarrow{\mu \otimes 1_M} C \otimes C \otimes M \xrightarrow{1_C \otimes f} C \otimes N \xrightarrow{g} L$$

Many classification problems in representation theory can be interpreted as problems in the classification of representations of bocses. For instance, representations of algebras, representations of posets, representations of bimodules. The theory of bocses was introduced by Roiter; this theory has some important applications.

In our situation C is finitely generated projective as a right and as a left module. We can take $B_r = Hom_A(C_A, A_A)$, this is an algebra over A, and we have an inclusion $A \to B_r$.

Let us now consider $p(B_r, A)$ the full subcategory of $modB_r$ with as objects the modules $B_r \otimes X$. Then, if r denotes the full subcategory of $\mathcal{R}$ consisting of those modules having finite dimension over k, $r \cong p(B_r, A) \subset mod(B_r)$.

Using a generalization of 3.2 one has the following result.

3.3 Theorem [8]: *The subcategory $p(B_r, A)$ has almost split sequences.*

The above result was proved by different methods by W.L. Burt and M.C. Butler [12] using 3.2. directly.

4. Stable equivalence.

As we mentioned before, tr is a duality between the categories $\underline{mod}\Lambda$ and $\underline{mod}\Lambda^{op}$ where $\underline{mod}\Lambda$ is defined as follows: objects of $\underline{mod}\Lambda$ = objects of $mod\Lambda$. Now if M and N are in $mod\Lambda$ we denote by $P(M, N)$ the set of morphisms which can be factorized through projectives. Then

$$\underline{Hom}_\Lambda(M, N) = Hom_\Lambda(M, N)/P(M, N)$$

4.1 Definition: The algebras Λ_1 and Λ_2 are said to be stably equivalent if $\underline{mod}\Lambda_1 \cong \underline{mod}\Lambda_2$.

The above defintion arises naturally in the theory of finite groups. One can ask what properties do two stably equivalent algebras have in common. In this direction we have the following conjecture.

4.2 Conjecture (Auslander-Reiten): *If Λ_1 and Λ_2 are stably equivalent then Λ_1 and Λ_2 have the same number of simples non projectives.*

This problem has attracted the attention of R. Martínez. He had many discussions with Maurice on this subject.

One of his first results was the characterization of algebras which are stably equivalent to $l-$hereditary algebras, generalizing previous results of M. Platzeck [20].

The ideas of this work were the basis for his work [18] in which each algebra Λ has an associated selfinjective algebra $\Sigma(\Lambda)$.

4.3 Theorem: *With the above notation Λ_1 and Λ_2 are stably equivalent if and only if $\Sigma(\Lambda_1)$ and $\Sigma(\Lambda_2)$ are stably equivalent.*

Using 4.3 the Auslander-Reiten conjecture is reduced to the case of self-injective algebras.

The conjecture 4.2 was solved in 1985.

4.4 Theorem: *If Λ_1 and Λ_2 are stably equivalent and of finite representation type then they have the same number of non projective simples.*

5. The influence of M. Auslander in the development of representation theory in Mexico.

After Maurice's second visit to Mexico, he visited our country on many more occasions. For instance he gave a lecture in Mazatlan during the meeting of the Mexican Mathematical Society. Later on he participated in the two meetings of the Latinoamerican School of Mathematics which were held in Mexico. He participated in the two international congress of representation theory organized in Mexico, the first in 1980 and the second in 1994.

Although we had different points of view, Maurice's influence was very important in the development of our group. This influence was not only through the suggestion of specific mathematical problems but through more general ideas of how to look at mathematics.

The ideas developed by R. Martínez and the author under the influence of Maurice were an important source of inspiration for the further work of Francisco Larrión and Leonardo Salmerón on simply connected algebras [15] and the work of R. Martínez and José Antonio de la Peña on coverings [19].

References

[1] M. Auslander and M.I. Platzeck. *Representation theory of hereditary artin algebras.* Lectures Notes in pure and applied mathematics, 37, Marcel Dekker, New York and Bassel (1978) 389-424.

[2] M.A. Auslander and I. Reiten. *Representations theory of artin algebras III.* Comm. in Algebra 3(1975) 239-294.

[3] M. Auslander and I. Reiten. *Representations theory of artin algebras V.* Comm. in Algebra 5(1977) 519-554.

[4] M. Auslander and S.O. Smalo. *Almost split sequences in subcategories.* J. Algebra 69 (1981) 426-454.

[5] R. Bautista. *Irreducible morphisms and the radical of a category.* An. Inst. Mat. 22, Univ. Nac. Aut. Mex. (1982) 83-135.

[6] R. Bautista. *On algebras close to hereditary artin algebras.* An. Inst. Mat. 21, Univ. Nac. Aut. Mex. (1981) 21-104.

[7] R. Bautista and S. Brenner. *Replication numbers for non-Dynkin sec-*

tional subgraphs in finite Auslander-Reiten quivers and some properties of Weyl roots. Proc. London Math. Soc. 43 (1983) 429-462.

[8] R. Bautista and M. Kleiner. *Almost split sequences for relatively projective modules.* J. Algebra, 135 (1) (1990) 19-56.

[9] R. Bautista and R. Martínez. *Representations of partially ordered sets and 1-Gorenstein algebras.* Proc. Conf. on Ring Theory. Antwerp (1978) Marcel Dekker, (1979) 385-433.

[10] I.N. Bernstein, I.M. Gelfand and V.A. Ponomarev. *Coxeter functors and Gabriel's theorem.* Usp. Mat. Nauk 28(1973) 19-33 Transl. Russ. Math. Surv. 28(1973) 17-32

[11] S. Brenner and M.C.R. Butler. *The equivalence of certain functors occurring in the representation theory of artin algebras and species.* J. London Math. Soc. 14(1976) 183-187.

[12] W.L. Burt and M.C. Butler. *Almost split sequences for bocses.* Preprint U. Liverpool (1990).

[13] V. Dlab and C.M. Ringel. *Indecomposable representations of graphs and algebras.* Memories Amer. Math. Soc. 173 (1976).

[14] P. Gabriel. *Representations indecomposables des ensembles ordonnes.* Seminaire P. Dubreil, Paris (1972-73) 301-304.

[15] F. Larrión and L. Salmerón. *On Auslander-Reiten quivers without oriented cycles.* Bull. London Math. Soc. 16 (1984) 47-51.

[16] M. Loupias. *Representations indecomposables des ensembles ordonnes finis.* These. Universite Francois Rabelais de Tours (1975).

[17] R. Martínez-Villa. *Almost projective modules and almost split sequences with indecomposable middle term.* Comm. in Algebra 8(1990) 1123-1150.

[18] R. Martínez-Villa. *Algebras stably equivalent to l-hereditary* Lectures Notes in Math. Springer 832 (1979) 396-431.

[19] R. Martínez-Villa and J. A. de la Peña. *Multiplicative basis for algebras whose universal cover has no oriented cycles.*

[20] M.I. Platzeck. *On algebras stably equivalent to an hereditary artin algebra.* Canadian J. Math. 30 (1978) 817-829.

INTERTWINED WITH MAURICE

by David A. Buchsbaum

What I would eventually like to do in this talk is describe some recent, if fragmentary, results on intertwining numbers. But given this rather special occasion, I thought I'd indulge in a bit of reminiscence and at the same time trace some of the twine that connects my present work with the spirit of the work that Maurice and I did so many years ago.

There are many bonds that interlace all of us here today. Of course you know that Maurice spent a good part of his middle and late life on representation theory of Artin algebras, but you may not realize that Emil Artin played a fundamental role in the very early mathematical lives of Maurice and me. When Maurice and I finished our theses at Columbia in 1953, Maurice went to Chicago and I went to Princeton. At that point, Maurice was still very much interested in group cohomology (he hated categories!), and I was puzzling over the implications of homological algebra to commutative ring theory, a puzzlement brought on by the Cartan-Eilenberg proof of the *Hilbert Syzygy Theorem*.

Although I had been invited to Princeton largely through the efforts of Steenrod, Emil Artin was kind enough to take a very young and ignorant new instructor seriously. I can't remember when in the 1953-54 academic year the topic came up, but it was in conversations with Artin that I first became aware of the open question: If R is regular, and P a prime ideal, is R_P regular? At that time, it was already clear that the global dimension of a regular local ring was finite; so to a naive optimist, thoroughly saturated with the latest in homological algebra, it was trivial to observe that if finite global dimension were characteristic of regular local rings, then simple localization arguments would guarantee a positive answer to that question. With some suggestions from Artin about sources that would help me to see the relationship between what I knew and what I wanted to know (e.g. Macaulay, Gröbner and I.S. Cohen), I got busy reading about the *Hauptidealklasse*, structure theorems, etc. With help and encouragement from E. Snapper (whom Artin had invited to Princeton to teach algebraic geometry while he was on leave), I succeeded in showing that the local rings of the cusp, the ordinary double point, and (to have integral closure and singularity simultaneously) the quadratic cone, all had infinite global dimension. (I say "with help from Snapper". You have to realize that at that point in my education, I barely knew what the equations for those varieties were!) Armed with this, and a "homologized" proof of the theorem in Gröbner that $hd_R(R/(I,x)) = 1 + hd_R(R/I)$ if x is not a zero divisor for R/I (I had essentially used a mapping cone argument--clearly a favorite sport of mine), I had a long conversation with Maurice (on the train to an AMS meeting at Rochester, NY) about undertaking the program to characterize regular local rings as those having finite global dimension. Needless to say, Maurice didn't get too enthusiastic until he had taken my puny effort in relation to $R/(I,x)$ and generalized it to the form we all know today about $hd_R M/xM = 1 + hd_R M$ if x is regular on M. (You all know that he was never one to hop onto a bandwagon; if the mathematics didn't speak to him, he'd have little or nothing to do with it. Fortunately he had already turned his interest to general homological algebra; his theorem that the global dimension of any ring is bounded by the projective dimensions of its cyclic modules was already in rough draft.) The important thing is that Maurice not only generalized the result, but used more elegant methods to prove it (methods that we now regard as purely elementary, but were still novel then). At that point he too became convinced that Ext and Tor had a real place in commutative algebra, and our collaboration began. (I should add here that it was Snapper who, after seeing that Maurice and I had been able to make progress on the "localization problem", mentioned to me that the problem of factoriality in regular local rings was still unsolved. As in the localization case, the theorem was known for geometric local rings, but not for unequal characteristic.)

There were very few times that Maurice and I actually got together physically to do our work. We mostly corresponded (long-distance phone calls were used in those days only to announce births and deaths), since neither of us could easily afford the cost of travel (the NSF had barely started to operate; our collaboration began in pre-Sputnik days). It was during one get-together that our theorem on the sum of the codimension and projective dimension of a module equaling the codimension of the ring was generated. The dynamic was typical of most of our work together: naive optimism confronted by deep scepticism and a final resolution. We also got together on our *Codimension and Multiplicity* paper; I visited with him for about a week. But even there, since he was an early riser and I a late one, we only overlapped for a few hours during the middle of the day and early evening, so our heated exchanges were of relatively short duration. He would drift off to bed while I continued working into the night, and I would leave him notes indicating where I had got to so that he could resume work when he awoke early the next day. We did manage to get that paper written, despite our spending so much time together on it!

You're all familiar with most of what we did together, so I'll just highlight a few facts that had a strong effect on my mathematics during this period. I've indicated that the *Hauptidealklasse* was a fundamental idea to which I was exposed early in my education on local ring theory. This of course appealed to me because it was the basis of the proof of the *Syzygy Theorem* which had intrigued me as a graduate student. I was led, naturally, to a systematic study and use of the Koszul complex, and its application to codimension and multiplicity (the connection with multiplicity was inherent in the work of Gröbner). The rigidity of the complex (which led to Maurice's *Rigidity Conjecture*--we were absolutely unable to solve the general rigidity problem when the notion presented itself in our *Codimension and Multiplicity* paper) was clearly a powerful tool, and was a condition I sought to satisfy when I embarked on the program to generalize the Koszul construction.

The reasons for seeking such a generalization were varied. One obvious one was the generalized *Cohen-Macaulay Theorem*: the Koszul complex had proven successful as a way to prove the usual *C-M* theorem, and the generalized one, having to do with the height of ideals generated by the minors of a matrix, was begging for a similar approach. Then there was the fact that matrices and modules were pretty much the same thing; I was always fascinated (and still am) by the fact that the presentation matrix of a module presumably holds all the information about that module. If that is the case, exploring the matrix systematically should give some additional techniques for studying arbitrary modules of finite type. (Of course, since a matrix and its transpose hold the same information, ...) Then, too, singular loci are determinantal, and I felt that the study of singularities would be enhanced by a good algebraic and homological treatment of minors. In conversations with Don Spencer, it became clear that overdetermined systems of PDEs might lend themselves to this sort of approach and finally, I thought that multiplicity theory, defined a la the Koszul complex technique, could be generalized to a larger category (this did lead to what is now known as the *Buchsbaum-Rim multiplicity*). I should add that I am a failed analyst and geometer; I have always been fascinated by analytic (mostly PDE) questions, and have frequently tried to do something with a geometric panache. As you know, I have never even gotten close, but my intentions have been honorable. In any event, I struggled for a long time with the problem of finding a generalized Koszul complex (Eagon and Northcott got in early with their minimal resolution of the ideal of maximal minors), did succeed in finding a family associated to the ideal of maximal minors (this included a complex which could be used as a resolution of a torsion module, i.e. the cokernel of the map associated to the matrix) and, with Rim, proceeded to develop this more general tool. To keep Maurice in the picture I should say that until I had proven that the complexes I had defined were "rigid", I couldn't be satisfied that I had the proper generalization of the Koszul complex. (The rigidity of these complexes implies an elementary but surprising fact about, even, the 2x2 minors of a 3x2 matrix over a local ring; write it down and see for yourselves.)

Although Maurice and I had diverged, we thought that we might be able to resume working together when I spoke to him about the lifting problem of Grothendieck (our joint interest, of course, was its application to intersection multiplicities). Maurice had some interesting obstruction cocycles at hand which I thought we might apply by a successive approximation technique to the lifting problem. Unfortunately, neither of us could make it work, and I, in the meantime, had realized that determinantal techniques might be gainfully applied to this problem. Maurice didn't like determinants, so our renewed collaboration aborted soon after its resumption. Let me review briefly why determinantal considerations enter the picture:

Suppose $R = S/(x)$, where R and S are local rings, and x is a regular element of S. Let M be a finitely generated R-module, and let

F: $\qquad \cdots \to F_k \to F_{k-1} \to \cdots \to F_2 \to F_1 \to F_0 \to M \to 0$

be a free R-resolution of M. By choosing bases for the free modules of the resolution, the maps may be described by matrices. Now let

$\overline{\mathbf{F}}$: $\qquad \cdots \to \overline{F}_k \to \overline{F}_{k-1} \to \cdots \to \overline{F}_2 \to \overline{F}_1 \to \overline{F}_0$

be a "lifting" of the complex F. By this I mean that the barred modules are free of the same rank as the corresponding unbarred free modules, and the maps are matrices whose entries in S are representatives of the entries in the corresponding matrices over R.
In short,

$$\overline{\mathbf{F}} \,/\, x\overline{\mathbf{F}} \;=\; \mathbf{F}.$$

Since x is regular in S, we have the exact sequence:

(*) $\qquad\qquad 0 \to \overline{\mathbf{F}} \overset{x}{\longrightarrow} \overline{\mathbf{F}} \to \mathbf{F} \to 0$

which, if it were a sequence of complexes, would imply that the barred complex is acyclic, and its 0-dimensional homology would be a lifting of M.

Thus, the problem is to lift a free resolution to a free complex and, in the case of a module, M, of projective dimension 0 or 1, this is clearly not a problem. The first case, then, is that of $pd_R(M) = 2$, and I first considered cyclic modules $M = R/I$. For this situation, I had to use a very special case of my family of complexes to show that I is essentially "determinantal". (I spoke about this in Rome and Genoa, and, when I returned home, showed the proof to Szpiro who, with Peskine, was visiting Maurice that year at Brandeis. In Genoa, David Rees had told me that he'd had a student by the name of Burch who had also recently, for very different reasons, proven much the same theorem. Neither Szpiro nor I had previously been aware of such a result [in fact, Peskine and Szpiro included a slightly improved version of my proof in a *Comptes Rendus* note. They also later provided a counter-example to the general "lifting problem"]. This turned out to be the well-known *Hilbert-Burch Theorem*, the connection with Hilbert having been discovered by Kaplansky, and Burch being the graduate student of whom Rees had spoken. It also turned out that a number of people had come up with the same result, and all for very different reasons.) The fact that such an ideal is determinantal made it possible to lift its resolution "by the tail", and thus show that it itself is liftable.

With this type of result and envisioned potential application, I thought that it should be possible to use determinantal information to analyze finite free resolutions. Fortunately for me, David Eisenbud arrived on the scene at Brandeis, and our very fruitful collaboration began. (I should admit that I had also become interested in D-modules and their applications--further evidence of my desire to do something in analysis--but Eisenbud's enthusiasm for the program I mentioned, and his criticism of my theretofore inelegant outline of it, seduced me from the unfamiliar back to familiar ground.) It became clear pretty soon after we started to work that the results of Maurice and Mark Bridger on Stable Module Theory had to be related to some of what we were doing. Of course we asked Maurice if he had any clues and, after a little thought, he characteristically said he didn't think so. I say "characteristically" because what it really meant was that he was busy doing something else, and therefore had forgotten the details of *Stable Module Theory*. (Maurice told me not too long ago that it was Idun Reiten who said of him that the only mathematics he knew was the mathematics that he had done himself and had not yet forgotten. This is probably true to a greater or lesser extent of any of us, but with Maurice it was amusingly, sometimes exasperatingly, extreme.) When Eisenbud and I did finally get our structure theorems for finite free resolutions, Maurice agreed that there was, after all, a connection. (For the issue of *Communications in Algebra* that was dedicated to his sixtieth birthday, I couldn't resist submitting a little article that was connected with these structure theorems and which related directly to part of the Auslander-Bridger work.)

By the late 70s, it was clear that Maurice's and my close collaboration was definitely over, although we of course still had a great many exchanges. Maurice would generally come into my office or stop me in the corridor and say that he wanted to check to see if a certain result was well-known. This was his way of saying that he had just come up with what he thought was a fascinating result, and he wanted to show it to me. In a way, our roles from our graduate school days had become reversed: in the old days, when I had been working on categories, he was the one who would ask for a concrete application of anything I'd show him. Now, he would come up with some categorical existence theorem, and it would be I who would ask what it looked like in such and such a case. We had both moved off into representation theory, but in very different ways. Most of you know only too well the directions he moved in. For the rest of this talk, I'll try to indicate how, in disentwining from Maurice, I became entangled with, among other things, *intertwining numbers*.

The introduction by Lascoux of classical representation-theoretic techniques to obtain resolutions of determinantal ideals was a tour-de-force. Since these classical techniques required the assumption of caracteristic zero (meaning that the ground ring contained the rationals), it was natural to ask whether the extension to arbitrary characteristic was possible and, if so, whether one could reproduce these resolutions in a characteristic-free way. The work of Carter and Lusztig ([C-L]) on representations over fields of arbitrary characteristic was in the literature, and the later work of Towber ([T]) was also available. In each case there were deficiencies from the point of view of the applications I had in mind; mainly I had to have a larger category of representations to deal with. That is, in order to make parallels between, say, the classical decomposition results and the characteristic-free ones, some combinatorial techniques had to be replaced by exact and spectral sequences. Obviously, to have such sequences, one must have a large enough category of modules at hand. So, with K. Akin and J. Weyman [A-B-W] we developed the notions of *Schur* and *Weyl modules* associated with arbitrary *shape matrices* (originally these were called *Schur* and *Coschur functors*). These reconstructed the usual representations if the shape matrix were that of a partition or a skew-partition, but we also had new shapes which had never been treated in the classical theory and which are nevertheless essential; the first of these new shapes emerged as the kernel of a very natural surjection between skew-partitions, and others as kernels of surjections between these, etc. At the cost of a certain amount of redundancy, let me quickly run through some basic definitions.

A *shape matrix* is an sxt matrix:

$$A = \left(a_{ij} \right), \quad \text{with } a_{ij} = 0 \text{ or } 1.$$

Denote by a_i the row sum of the ith row of the matrix, and by b_j the column sum of its jth column.

Given a free module, F, over the commutative ring, R, the *Schur* and *Weyl maps* associated to the shape matrix and the module F are defined as the following composite maps:

$$\Lambda^{a_1} F \otimes \cdots \otimes \Lambda^{a_s} F \to \begin{array}{c} \Lambda^{a_{11}} F \otimes \cdots \otimes \Lambda^{a_{1t}} F \otimes \\ \Lambda^{a_{21}} F \otimes \cdots \otimes \Lambda^{a_{2t}} F \otimes \\ \vdots \\ \Lambda^{a_{s1}} F \otimes \cdots \otimes \Lambda^{a_{st}} F \end{array} \xrightarrow{\quad} \begin{array}{c} S_{a_{11}} F \otimes \cdots \otimes S_{a_{1t}} F \otimes \\ S_{a_{21}} F \otimes \cdots \otimes S_{a_{2t}} F \otimes \\ \vdots \\ S_{a_{s1}} F \otimes \cdots \otimes S_{a_{st}} F \end{array} \to S_{b_1} F \otimes \cdots \otimes S_{b_t} F ;$$

$$D_{a_1} F \otimes \cdots \otimes D_{a_s} F \to \begin{array}{c} D_{a_{11}} F \otimes \cdots \otimes D_{a_{1t}} F \otimes \\ D_{a_{21}} F \otimes \cdots \otimes D_{a_{2t}} F \otimes \\ \vdots \\ D_{a_{s1}} F \otimes \cdots \otimes D_{a_{st}} F \end{array} \xrightarrow{\quad} \begin{array}{c} \Lambda^{a_{11}} F \otimes \cdots \otimes \Lambda^{a_{1t}} F \otimes \\ \Lambda^{a_{21}} F \otimes \cdots \otimes \Lambda^{a_{2t}} F \otimes \\ \vdots \\ \Lambda^{a_{s1}} F \otimes \cdots \otimes \Lambda^{a_{st}} F \end{array} \to \Lambda^{b_1} F \otimes \cdots \otimes \Lambda^{b_t} F ,$$

where the left maps are diagonalizations as indicated, the middle maps are the obvious identifications (due to the fact that a_{ij} are all 0 or 1), and the right maps are the multiplication maps down the columns.

The image of the first map is called the *Schur module of F of shape A*, and the image of the second, the *Weyl module of F of shape A*, denoted $L_A(F)$ and $K_A(F)$

respectively. One of the fundamental theorems proven about these modules, when A is the shape matrix of a skew-partition, is the *Standard Basis Theorem*.

(Recall that the matrix A is called the *shape matrix* of the skew-partition λ / μ if

$$a_{ij} = 0 \ for \ j = 1, \cdots, \mu_i;$$

$$a_{ij} = 1 \ for \ \mu_i + 1 \le j \le \lambda_i.$$

This corresponds to the shape:

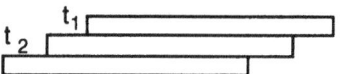

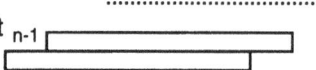

where each row has $\lambda_i - \mu_i$ boxes for i = 1, ..., n, and $t_i = \mu_i - \mu_{i+1}$ for $i = 1, ..., n\text{-}1$. The corresponding matrix has the form:

$$
\begin{pmatrix}
\underbrace{00 \quad \cdots \quad 0001}_{\mu_1} \underbrace{1 \quad \cdots \quad 1111}_{\lambda_1 - \mu_1} \\
\underbrace{00 \quad \cdots \quad 0011}_{\mu_2} \underbrace{\cdots \quad 11100}_{\lambda_2 - \mu_2} \\
\vdots \qquad\qquad \vdots \\
\underbrace{00 \quad \cdots \quad 011}_{\mu_{n-1}} \underbrace{\cdots \quad 110000}_{\lambda_{n-1} - \mu_{n-1}} \\
\underbrace{0 \quad \cdots \quad 01}_{\mu_n} \underbrace{\cdots \quad 10000000}_{\lambda_n - \mu_n}
\end{pmatrix}.
$$

We then usually write $L_{\lambda/\mu}(F), K_{\lambda/\mu}(F)$ instead of $L_A(F)$ and $K_A(F)$.)

The *Standard Basis Theorem* says that for skew-partitions, the modules $L_{\lambda/\mu}(F)$ and $K_{\lambda/\mu}(F)$ are universally free, and that their bases can be described as the set of standard tableaux in terms of a given basis of F. The universality of these modules implies that they are obtained from their integral forms by extension of base ring (or reduction of base ring, in the case of modular representations).

A special case of a more general theorem found in [A-B1] says the following:

Let M and N be integral polynomial representations of $Gl(F)$ of degree r, let p be a prime, and let A_r be the integral Schur algebra of degree r. Denote by $\overline{X}$ the $\mathbf{Z}/(p)$ - module (- algebra) $X \otimes_{\mathbf{Z}} \mathbf{Z}/(p)$. Then we have the exact sequence:

$$0 \to Ext^i_{A_r}(M,N) \otimes \mathbf{Z}/(p) \to Ext^i_{\overline{A}_r}(\overline{M},\overline{N}) \to Tor^1_{\mathbf{Z}}(Ext^{i+1}_{A_r}(M,N),\mathbf{Z}/(p)) \to 0.$$

Suppose now that λ is a partition: $\lambda = (\lambda_1, \cdots, \lambda_n)$, that d is a positive integer, and that μ is the partition obtained from λ by taking d boxes away from some row of λ and attaching them to some higher row (assuming that we still obtain a partition), i.e.

$$\mu = (\lambda_1, \cdots, \lambda_k + d, \cdots, \lambda_{k+j} - d, \cdots, \lambda_n).$$

A standard question in modular representation theory is: What is the $\mathbf{Z}/(p)$-dimension of the $\mathbf{Z}/(p)$-vector space $Ext^i_{\overline{A}_r}(\overline{K}_\lambda, \overline{K}_\mu)$, where the notation is as indicated above. These numbers are called *intertwining numbers*. From the exact sequence above, we see that it suffices to calculate the integral Ext groups, since the modular ones are simply the p-torsion part of one integral Ext plus the reduction modulo p of another.

A fairly straightforward argument shows that it suffices to consider the case where we take d boxes from the last row and attach them to the first one. As you might imagine, this is not the heart of the difficulty. Of course, straight homological algebra tells us that we merely have to find a projective resolution, $\mathbf{P}$, of K_λ over A_r, and then simply calculate the cohomology of $Hom_{A_r}(\mathbf{P}, K_\mu)$.

In [A-B1], Akin and I proved that K_λ admits a finite projective resolution whose terms are direct sums of tensor products of divided powers (suitable tensor products of divided powers are A_r-projective). Moreover, we showed that, if $\alpha = (\alpha_1, \cdots, \alpha_n)$ is an integral weight of degree r, then $Hom_{A_r}(D_{\alpha_1} \otimes \cdots \otimes D_{\alpha_n}, N)$ is the weight submodule of the representation N corresponding to the weight α. Therefore, in our case with $N = K_\mu$, the calculation of $Hom_{A_r}(\mathbf{P}, K_\mu)$ comes down to the calculation of certain weight submodules of K_μ. That this is computable is due to the fact that the weight submodule of K_μ is the free abelian group generated by the standard tableaux of shape μ and content α.

Now we know that $Hom_{A_r}(\mathbf{P}, K_\mu)$ is a complex of free abelian groups, and that its cohomology is torsion (since over the rationals the cohomology is zero). If we write down the maps in $Hom_{A_r}(\mathbf{P}, K_\mu)$ as integral matrices, elementary arguments show us that the cohomology groups are determined by the invariant factors of those matrices (the non-zero invariant factors). To calculate those matrices, we must know the maps in the projective resolution $\mathbf{P}$. And so I've returned, as so often in the past, to the problem of finding some explicit form of projective resolutions, this time associated to Weyl modules.

In [A-B1], Akin and I wrote down explicit projective resolutions of two-rowed skew-partitions, and in [B-R1], Rota and I made use of letter-place techniques to define a splitting homotopy for these resolutions. Rota and I are now working to describe explicitly the resolutions of n-rowed shapes, and have succeeded in finding exactly what the terms of these resolutions are. As yet, we have not found the boundary maps.(We've made some progress on a class of three-rowed shapes, for which the boundary maps can be described [B-R2].) I won't go into detail on this subject, but I should say that the extended class of shapes that arose in my work with Akin ([A-B1]) plays a key role in this problem (I can even describe the terms in their resolutions). Since these shapes cannot be avoided in the characteristic-free representation theory, their study seems to be of prime importance. But more about that another time.

However, we can momentarily avoid the problem of exhibiting these resolutions if we're willing to just study the intertwining numbers for $i = 0$, that is if we just want to study the dimension of $Hom_{\overline{\Lambda}_r}(\overline{K}_\lambda, \overline{K}_\mu)$. For in this case, as already observed, we need only calculate the p-torsion of $Ext^1_{\Lambda_r}(K_\lambda, K_\mu)$ (the corresponding Hom group is zero), and from the remarks I made about the nature of the cohomology of $Hom_{\Lambda_r}(\mathbf{P}, K_\mu)$, it suffices to know a presentation of K_λ. And this we do know, thanks to the proof of the *Standard Basis Theorem*. For two-rowed shapes and arbitrary d, the problem was solved many years ago by Akin and me ([A-B2]). In 1987 I spoke about this problem for three-rowed shapes at MSRI ([A-B2]), and wrote down the integral matrix whose invariant factors have to be computed. That same summer, I pointed out to D. Flores that in the three-rowed case it was not yet even proven (although conjectured) that the Ext group was cyclic. In 1991, she proved that it was indeed cyclic [F], and we then started to look for the highest invariant factor. Akin and I had made a conjecture, or rather a guess, about it (because we had calculated the case $d = 2$), but in 1988, by using a computer at the Politecnico di Torino, I found that that guess was wrong, and passed on a modified guess to Flores. She worked on the problem for a while, enough to find that the last guess was wrong too but, since the matrices are rather large (you'll soon see), it was getting more and more difficult to get insight into the computations. Finally, around a year ago, a graduate student of Eisenbud's (who has since finished his degree) named Michael Johnson wrote out a subprogram in Maple which enabled us to check some larger examples (the size of the problem depends upon d), and to get some idea as to the arithmetic that was going on. Finally last fall, we hit on the pattern that seemed to make sense, and by December we had a proof that our guess was correct. I'll end by outlining very broadly some of what is involved in the proof. A detailed account of it will be published elsewhere [B-F].

We start with a three-rowed partition $\lambda = (\lambda_1, \lambda_2, \lambda_3)$, an integer $d \leq \lambda_3$, and we define the partition μ to be $(\lambda_1 + d, \lambda_2, \lambda_3 - d)$. If we define $s_1 = \lambda_1 - \lambda_2$, and $s_2 = \lambda_2 - \lambda_3$, calculations described in [AB2] tell us that $Ext^1_{\Lambda_r}(K_\lambda, K_\mu)$ is $\mathbf{Z}/(\delta)$, where δ is the highest invariant factor of the matrix:

$$M(s_1, s_2, d) = \left\langle A_d | B_d \left| \begin{matrix} 0 \cdots 0 \\ M(s_1, s_2 + 1, d - 1) \end{matrix} \right. \right\rangle$$

where

$$A_d = \begin{pmatrix} s_1+d+1 & -\binom{s_1+d+1}{2} & \cdots & (-1)^{d-1}\binom{s_1+d+1}{d} \\ -\binom{d}{1} & 0 & \cdots & 0 \\ 0 & \binom{d}{2} & \cdots & 0 \\ \vdots & \vdots & \vdots & \vdots \\ 0 & 0 & \cdots & (-1)^d\binom{d}{d} \end{pmatrix};$$

$$B_d = \begin{pmatrix} s_2+1 & -\binom{s_2+2}{2} & \cdots & (-1)^{d-1}\binom{s_2+d}{d} \\ \binom{d}{1} & 0 & \cdots & 0 \\ 0 & \binom{d}{2} & \cdots & 0 \\ \vdots & \vdots & \vdots & \vdots \\ 0 & 0 & \cdots & \binom{d}{d} \end{pmatrix}.$$

By adding to the jth column of B_d the jth column of A_d multiplied by $(-1)^{j-1}$, the block B_d takes the form:

$$B_d' = \begin{pmatrix} \alpha & H_2 & \cdots & H_d \\ 0 & 0 & \cdots & 0 \\ \vdots & \vdots & \cdots & \vdots \\ 0 & 0 & \cdots & 0 \end{pmatrix},$$

where $\quad H_j = \binom{s_1+d+1}{j} + (-1)^{j-1}\binom{s_2+j}{j}.$

Since $H_j = \sum\limits_{l=1}^{j}\binom{\alpha}{l}\binom{-(s_2+1)}{j-l}$, where $\alpha = s_1 + s_2 + d + 2$, we reduce B_d', by column transformations, to:

$$B_d^{''} = \begin{pmatrix} \alpha & \binom{\alpha+1}{2} & \cdots & \binom{\alpha+d+1}{d} \\ 0 & 0 & \cdots & 0 \\ \vdots & \vdots & \cdots & \vdots \\ 0 & 0 & \cdots & 0 \end{pmatrix}.$$

Next, since

$$\alpha_d = \frac{\alpha}{\gcd(\alpha, lcm\{1,\cdots,d\}} = \gcd\left\{\alpha, \binom{\alpha+1}{2}, \cdots, \binom{\alpha+d-1}{d}\right\},$$

we may reduce the matrix $B_d^{''}$, by column transformations, to the matrix:

$$B_d^{'''} = \begin{pmatrix} \alpha_d & 0 & \cdots & 0 \\ 0 & 0 & \cdots & 0 \\ \vdots & \vdots & \vdots & \vdots \\ 0 & 0 & & 0 \end{pmatrix}.$$

If to the last column of each block A_{d-j}, $j = 0,...,d-2$, we add the last column of A_{d-j-1}, each block A_{d-j} takes the form:

$$A_{d-j}' = \begin{pmatrix} (s_1+d-j+1) & -\binom{s_1+d-j+1}{2} & \cdots & (-1)^{d-j-1}\binom{s_1+d-j+1}{d-j} \\ -(d-j) & 0 & \cdots & (-1)^{d-j}\binom{s_1+d-j}{d-j-1} \\ 0 & \binom{d-j}{2} & \cdots & 0 \\ \vdots & \vdots & \vdots & \vdots \\ 0 & 0 & \cdots & 0 \end{pmatrix}.$$

We should note that when $j = d-2$,

$$A_2' = \begin{pmatrix} (s_1+3) & -\binom{s_1+3}{2} \\ -2 & (s_1+2) \end{pmatrix}.$$

This leads us to make the following slightly different definitions.

Define:

$$N(\alpha,s,2) = \begin{pmatrix} \alpha_2 & 0 & s+3 & -\binom{s+3}{2} \\ 0 & \alpha & -2 & s+2 \end{pmatrix}$$

and, inductively,

$$N(\alpha,s,d) = \begin{pmatrix} \alpha_d & s+d+1 & -\binom{s+d+1}{2} & \cdots & (-1)^{d-1}\binom{s+d+1}{d} & 0 & 0 & \cdots & 0 \\ 0 & -\binom{d}{1} & 0 & \cdots & (-1)^{d-2}\binom{s+d}{d-1} & N(\alpha,s,d-1) \\ \vdots & & \binom{d}{2} & & & 0 \\ 0 & & \vdots & & & \vdots \\ 0 & & 0 & \cdots (-1)^{d-1}\binom{d}{d-1} & & 0 \end{pmatrix}$$

where α and d are positive integers, s is a non-negative integer, and $\alpha > s+d+1$.

It is easy to see that for $d \geq 2$, $N(\alpha,s,d)$ is a $d \times \left\{\binom{d+2}{2} - 2\right\}$ matrix.

The main result that D. Flores and I have at this point is:

THEOREM: Let $\delta(\alpha,s,d)$ denote the highest invariant factor of the matrix $N(\alpha,s,d)$ above. Then

$$\delta(\alpha,s,d) = \frac{\alpha}{\gamma(\alpha,s+d+1,d)} \prod_{k=1}^{\left[\frac{d-1}{2}\right]} \frac{\gamma(\alpha,s+d-k+1,k)}{\gamma(\alpha,s+d-k+1,d-k)},$$

where $\gamma(u,v,w) = \gcd\{u,v,lcm\{1,\cdots,w\}\}$.

As you can see, the proof is largely an organization of these matrices into blocks that we can handle. This accounts for the next bit of notation:

For positive d and non-negative s (integers), define:

$$A'(s,d) = \begin{pmatrix} s+d+1 & -\binom{s+d+1}{2} & \cdots & (-1)^{d-2}\binom{s+d+1}{d-1} & (-1)^{d-1}\binom{s+d+1}{d} \\ -\binom{d}{1} & 0 & \cdots & 0 & (-1)^{d-2}\binom{s+d}{d-1} \\ 0 & \binom{d}{2} & \cdots & 0 & 0 \\ \vdots & \vdots & \vdots & \vdots & \vdots \\ 0 & 0 & 0 & (-1)^{d-1}\binom{d}{d-1} & 0 \end{pmatrix},$$

and set

$$A(s,d) = \begin{pmatrix} A'(s,d) & 0 & \cdots & 0 \\ \vdots & A'(s,d-1) & \cdots & 0 \\ \vdots & \vdots & \cdots & \vdots \\ 0 & 0 & \cdots & A'(s,2) \end{pmatrix},$$

and

$$B(\alpha,d) = \begin{pmatrix} \alpha_d & 0 & \cdots & 0 \\ & \alpha_{d-1} & \cdots & 0 \\ \vdots & \vdots & \vdots & \vdots \\ 0 & 0 & \cdots & \alpha \end{pmatrix}.$$

By induction on d, $(d \geq 2)$ we show that the $d \times d$ minors of $A(s,d)$ are all zero. This latter is shown by using elementary column transformations in Q.

Once the above is proven, we know that the $d \times d$ minors of our matrix $N(\alpha, s, d)$ are very special. The rest of the argument is a detailed study of the p-divisibility properties of certain ones of these minors.

The proof as it now stands is not too transparent, and should be improved. The reason for this is not simply a matter of elegance, but necessity: in studying the n-rowed case (with R. Sanchez in addition to D. Flores) there are a number of recursions that one sees entering the picture, and matrices very similar to, but much larger than the ones here considered have to be dealt with.

You can see how far Maurice and I had diverged by the time we had both wandered into representation theory. I suppose one can say, though, that the common thread that we always held onto was the belief that more traditional mathematics was amenable to treatment by homological methods. When we were taking Sammy Eilenberg's course in Homological Algebra, Sammy would occasionally remark that "the universe is cohomological". He meant this always in the sense that cohomology is more "natural" than homology. I think that Maurice and I tended to carry the dictum one step further: the universe *is* cohomology. The textbook that we wrote together was a public sermon to that effect; we were going to show the world how classical algebra could have developed if homological algebra had been around since creation. For me, I know, the main part of my work has been to detect the "call for homology" in various parts of mathematics. The localization and factoriality problems in local ring theory were undertaken not so much because I was dying to know if they were true or not (after a while, of course, I *was* interested in the outcome), but because I felt that they were an area calling out for homological treatment. You see that also in my approach to characteristic-free representation theory. I believe that Maurice was impelled by the same credo. Although less obsessive than I in pursuing some problems to the very end, his early involvement in commutative ring theory (with me and with others), and certainly his unique categorical and split-exact approach to representation theory bear this out. This belief was of course always tempered by reality; if not we'd have been in serious trouble!

REFERENCES

[A-B1]   K. Akin and D. A. Buchsbaum, Characteristic-free representation theory of the general linear group, II. Homological considerations. *Adv. in Math.* V 72 No. 2 (1988) 171-210.

[A-B2]   K. Akin and D. A. Buchsbaum, Representations, resolutions and intertwining numbers. *Comm. Algebra, MSRI Publications, Springer* 1989, pp. 1-19.

[A-BW]   K. Akin, D. A. Buchsbaum and J. Weyman, Schur functors and Schur complexes. *Adv. in Math.* 44, No. 3 (1982), 207 - 278.

[B-F]   D. A. Buchsbaum and D. Flores de Chela, Intertwining numbers: the three-rowed case. *To appear in J. Algebra*

[B-R1]   D. A. Buchsbaum and G-C Rota, Projective Resolutions of Weyl Modules. *Proc. Natl. Acad. Sci. USA*, vol 90, pp 2448-2450, (March, 1993)

[B-R2]   D. A. Buchsbaum and G-C Rota, A New Construction in Homological Algebra. *Proc. Natl. Acad. Sci. USA*, vol 91, pp 4115-4119, (May,1994)

[C-L]   R.W. Carter and G. Lusztig, On the modular representations of the general linear and symmetric groups. *Math. Z.* 136 (1974), pp 193 - 242.

[F]   D. Flores de Chela On Intertwining Numbers. *J. Algebra 171 (1995), 631 - 653.*

[T]   J. Towber, Two new functors from modules to algebras. *J. Algebra 47 (1977), 80 - 109.*

Introduction to Koszul algebras

Edward L. Green *

Virginia Polytechnic Institute

and State Universtity

Blacksburg, Virgina USA

September 6, 1996

This paper is dedicated to the memory of Maurice Auslander,
whose love of mathematics will always be an inspiration

Abstract

In this paper we survey, without proof, the main structural results
of Koszul algebras. We also survey the extension of this theory to
semiperfect Noetherian rings. Applications to algebras of global dimensions 1 and 2 are discussed as well as applications to Auslander
algebras and to preprojective algebras.

1 Introduction and Definititons

Koszul algebras have played an importart role in commutative algebra and
algebraic topology [22, 16, 17, 3, 4, 24, 25]. Recently there have been important applications of noncommutative Koszul algebras to algebraic topology,
Lie theory and quantum groups [8, 7, 23, 5]. This paper begins with some
terminology and definitions in this section. In section 2, we survey the main
results about Koszul algebras. Definition of a Koszul algebra follows in this

1991 Mathematics Subject Classification. Primary 16W50; Secondary 16G20

*This paper has been written with support from the NSF

section. The proofs of the results in this section can be found in [14, 15].
In section 3, we survey the main results about Koszul modules stressing the
duality between the category of Koszul modules of a Koszul algebra and the
category of Koszul modules over the Yoneda algebra. Again, proofs may
be found in [14, 15]. The cited references, the concept of a Koszul algebra
and module is extended outside the graded case. The notion of quasi-Koszul
algebra is introduced. A definition is given later in this section. This extension is summarized in section 4. Section 5 surveys relations between Koszul
algebras and algebras of global dimensions 1 and 2. The Auslander algebra
is also introduced there and shown to be quasi-Koszul. The results up to this
point, in the generality described, was joint work with R. Martínez Villa.

Section 6 briefly introduces the notion of a Gröbner basis and shows who
these are related to Koszul algebras. Finally section 7 applies these ideas
to show that a certain class of generalized preprojective algebras are Koszul
algebras.

We study quotients of path algebras, a class of rings that include quotients
of free associative algebras and quotients of commutative polynomial rings.
The generality provided by path algebras, algebras which naturally occur in
the studying finite dimensional algebras, justifies studying the nonlocal case.

We briefly recall the definition of a path algebra and refer the interested
reader to [2] for further details. The notation introduced in this section will
be used throughout this paper.

Let Γ be a finite directed graph and K a fixed field. The *path algebra*, $K\Gamma$,
is defined to be the K-algebra having as K-basis the finite directed paths in
Γ. Thus, elements of $K\Gamma$ are finite K-linear combinations of paths. "Path"
will always mean directed walk in Γ in all that follows. We let Γ_0 denote the
vertex set of Γ and Γ_1 denote the arrow set of Γ. If $a \in \Gamma_1$ then we let $o(a)$
denote the origin vertex of a and $t(a)$ denote the terminus vertex. We will
sometimes write $a : o(a) \rightarrow t(a)$. If $v \in \Gamma_0$ we set $o(v) = v = t(v)$. Note that
the vertices of Γ are viewed as paths of length 0, where the length of a path
is the number of arrows occuring in the path. If $p = a_n \dots a_1$ is a path with
$a_i \in \Gamma_1$, then $o(a_i) = t(a_{i-1})$ for $i = 2, \dots, n$. The multiplicative structure is
given by linearly extending the following multiplication of paths. If p and q
are paths then we define $p \cdot q = \{ \begin{array}{ll} pq & \text{if } t(q) = o(p), \\ 0 & \text{otherwise.} \end{array}$ Note that if $v = o(p)$
or $v = t(q)$ then we define $p \cdot v = p$ and $v \cdot q = q$.

There is an alternative description for path algebras (see [2, 11]). If R
is a finite product of copies of K we may view R as a K-algebra where K
acts diagonally in R. If M is a finite K-dimensional R-R-bimodule then the
tensor algebra $T_R(M) = R \oplus M \oplus (M \otimes_R M) \oplus (\otimes_R^3 M) \oplus \cdots$ is isomorphic to a

path algebra. Conversely, if $K\Gamma$ is a path algebra, the subalgebra generated by the vertices, R, is isomorphic to a finite product of K's. Moveover, the vector space generated by the arrows, call it M, is an R-R-bimodule. Then $K\Gamma$ is isomorphic to the tensor algebra $T_R(M)$. It is easy to see that the free associative algebra in n noncommuting variables is isomorphic to the the path algebra $K\Gamma$ where Γ is a graph with exactly one vertex and n loops at the vertex.

We will view a path algebra as a positively $\mathbf{Z}$-graded algebra with the paths being homogeneous elements of degree equal to their lengths. We will call this grading the *length grading*. Viewed as a tensor algebra $T_R(M)$, this is the natural grading where $T_R(M)_n = \otimes_R^n M$. If I is a homogeneous ideal in $K\Gamma$, then the quotient ring, $K\Gamma/I$, inherits the positive $\mathbf{Z}$-grading. As further justification of studying path algebras, we have the following result. Suppose that $S = S_0 + S_1 + S_2 + \cdots$ is a positively $\mathbf{Z}$-graded K-algebra satisfying the following three conditions.

1. S_0 is isomorphic to a finite product of copies of the field K.

2. Each S_i is a finite dimensional K vector space.

3. S is generated in degrees 0 and 1.

The last condition means that $S_1 S_n = S_{n+1}$ for all $n \geq 1$. Under these assumptions, S is isomorphic to a graded quotient of a path algebra (with the length grading).

We now introduce more notation and conventions which will be used in the remainder of paper. Let J denote the two-sided ideal in $K\Gamma$ generated by the arrows of Γ. Note that J is the graded Jacobson radical of $K\Gamma$ with the length grading. We let I denote a two-sided ideal in $K\Gamma$ and we will always assume that $I \subseteq J^2$. Let $\Lambda = K\Gamma/I$ be the quotient algebra. We will assume that either I is a homogeneous ideal in the length grading or that Λ is a Noetherian semiperfect K-algebra. In the former case, we let $\mathbf{r}$ denote graded Jacobson radical of Λ; that is, if $\Lambda = \Lambda_0 + \Lambda_1 + \Lambda_2 + \cdots$ then $\mathbf{r} = \Lambda_1 + \Lambda_2 + \cdots$. In the latter case, we require that $\mathbf{r}$ is the Jacobson radical of Λ. We will call the first case, *the graded case* and the second case *the nongraded case*.

In the graded case, we will be interested in finitely generated graded modules and degree 0 homomorphisms; whereas, in the nongraded case, we will be interested in finitely generated modules and Λ-homomorphisms. In both cases, the assumptions on Λ imply the existence of minimal projective resolutions of modules. Of course, in the graded case, this will mean graded projective resolutions.

Let $E(\Lambda) = \coprod_{n \geq 0} \mathrm{Ext}_\Lambda^n(\Lambda/\mathbf{r}, \Lambda/\mathbf{r}) = \mathrm{Ext}_\Lambda^*(\Lambda/\mathbf{r}, \Lambda/\mathbf{r})$ be the Yoneda algebra of Λ. That is, $E(\Lambda)$ is given a multiplicative structure using the Yoneda product (see [19]). $E(\Lambda)$ has also been called the Ext-algebra of Λ and also the cohomology ring of Λ. We view $E(\Lambda)$ as a graded K-algebra via $E(\Lambda)_n = \mathrm{Ext}_\Lambda^n(\Lambda/\mathbf{r}, \Lambda/\mathbf{r})$. We say that Λ is a *quasi-Koszul* algebra if $E(\Lambda)$ is generated in degrees 0 and 1; that is, if $\mathrm{Ext}_\Lambda^1(\Lambda/\mathbf{r}, \Lambda/\mathbf{r}) \cdot \mathrm{Ext}_\Lambda^n(\Lambda/\mathbf{r}, \Lambda/\mathbf{r}) = \mathrm{Ext}_\Lambda^{n+1}(\Lambda/\mathbf{r}, \Lambda/\mathbf{r})$ for all $n \geq 0$. In the graded case, if $E(\Lambda)$ is generated in degrees 0 and 1 we say that $E(\Lambda)$ is a *Koszul algebra*.

2   Fundamental Theorems on Koszul Algebras

Throughout this section, we assume that $\Lambda = K\Gamma/I$ is a graded K-algebra with the length grading and that I is contained in J^2 where J is the ideal in $K\Gamma$ generated by the arrows. We say a graded Λ-module M is *generated in degree i* if $M_n = \Lambda_{n-i}M_i$ for all $n \in \mathbf{Z}$. We say a graded module M has a *linear resolution* if M is generated in degree 0 and if there is a graded projective Λ-resolution of M

$$\cdots \to P_n \xrightarrow{f_n} P_{n-1} \to \cdots \to P_1 \xrightarrow{f_1} P_0 \xrightarrow{f_0} M \to 0$$

such that for $n \geq 0$, P_n is generated in degree n. We get the first main result which provides an equivalent description of a Koszul algebra.

Theorem 2.1 *[14, Cor 3.4] [22, 8] Let $\Lambda = K\Gamma/I$ be as above. Then the following statements are equivalent:*

  1. *Λ is a Koszul algebra.*

  2. *As a graded Λ-module generated in degree 0, $\Lambda/\mathbf{r}$ has a linear resolution.*

□

The next result is also fundamental.

Theorem 2.2 *[14, Thm 6.1] [22, 8] If $\Lambda = K\Gamma/I$ is a Koszul algebra then the Yoneda algebra $E(\Lambda) = Ext_\Lambda^*(\Lambda/\mathbf{r}, \Lambda/\mathbf{r})$ is a Koszul algebra.* □

We remark that even if Λ is a commutative algebra, $E(\Lambda)$ is, in general, not commutative. Furthermore, even if Λ is finite dimensional, $E(\Lambda)$ need not be Noetherian. Thus, the setting of noncommutative rings which are not necessarily Noetherian naturally arises in the study of Koszul algebras.

We say that $\Lambda = K\Gamma/I$ is a *quadratic algebra* if I is generated linear combinations of paths of length 2. Note that Λ is quadratic if and only if I is homogeneous in the length grading and generated by elements of degree 2.

Proposition 2.3 *[14, Cor 7.3] If Λ is a Koszul algebra then Λ is a quadratic algebra.* $\square$

We remark that the converse is not true in general. For example, if Γ is the graph

$$
\begin{array}{ccc}
& a & c \\
\overset{v_1}{\circ} & \overset{\longrightarrow}{\underset{\longleftarrow}{}} & \overset{v_2}{\circ} & \overset{\longrightarrow}{\underset{\longleftarrow}{}} & \overset{v_3}{\circ} \\
& b & d
\end{array}
$$

and I is the ideal generated by $ba, ab + dc$, and cd in $K\Gamma$ then although I is quadratic, it is not hard to show that $K\Gamma/I$ is not a Koszul algebra.

The next result describes the algebra structure of the Yoneda algebra of a Koszul algebra. For this we need some further notation and definitions. We let Γ^{op} denote the *opposite quiver of* Γ. That is, $\Gamma_0^{op} = \Gamma_0$ and $\Gamma_1^o = \{a^o : v \to w \mid$ where $a : w \to v$ is an arrow in $\Gamma\}$. Let V_2 denote the vector space in $K\Gamma$ spanned by the paths of length 2 in Γ and let V_2^o denote the vector space in $K\Gamma^{op}$ spanned by the paths of length 2 in Γ^{op}. Let $x_1, \ldots, x_m$ be the K-basis of V_2 of paths of length 2. Similarly, let $x_1^o, \ldots, x_m^o$ be the corresponding basis of of V_2^o; that is, if $x_i = ab$ with $a, b \in \Gamma_1$ then $x_i^o = b^o a^o$. If W is a K-vector space we let W^*, the *dual space of* W, be $\mathrm{Hom}_K(W, K)$. We let $\{x_i^*\}$ denote the dual basis of V_2^* associated to the basis $\{x_i\}$ of V_2. Then we have a canonical bilinear form

$$
< , >: V_2 \times V_2^o \to K
$$

given by $< x_i, x_j^o >= x_j^*(x_i)$ where $x_j^* \in V_2^*$ is part of the dual basis $\{x_i^*\}$ associated to the basis $\{x_i\}$ of V_2. Finally, if Z is a subspace of V_2, then $Z^\perp = \{z^o \in V_2^o \mid < Z, z^o >= 0\}$.

We can now give the structure of the Yoneda algebra of a Koszul algebra.

Theorem 2.4 *[15, Thm 2.2] [17] Let $\Lambda = K\Gamma/I$ be a Koszul algebra and let $I_2 = I \cap V_2$. Then $E(\Lambda)$ is isomorphic to $K\Gamma^{op}/< I_2^\perp >$.* $\square$

Note that if Z is a subspace of V_2, then $(Z^\perp)^\perp = Z$. We get the following result.

Corollary 2.5 *[15, Thm 2.3] Let $\Lambda = K\Gamma/I$ be a Koszul algebra. Then $E(E(\Lambda))$ is isomorphic to Λ as graded algebras.* $\square$

We have a partial converse to this corollary.

Theorem 2.6 *[15, Thm 2.4] Let $\Lambda = K\Gamma/I$ be a graded algebra with the length grading. Then Λ is a Koszul algebra if and only if $E(E(\Lambda))$, the Yoneda algebra of the Yoneda algebra of Λ is isomorphic to Λ as graded algebras.* □

We give two final ring theoretic results about Koszul algebras.

Proposition 2.7 *[15, Cor 4.3] If $\Lambda = K\Gamma/I$ is a Koszul algebra then the opposite algebra of Λ, Λ^{op} is also a Koszul algebra.* □

This result can be interpretted as saying that Λ is a Koszul algebra if and only if $\Lambda/\mathbf{r}$ has a colinear injective resolution. The definition of a colinear injective resolution is left to the reader.

Theorem 2.8 *[15, Thm 3.7] If $\Lambda = K\Gamma/I$ and $\Lambda' = K\Gamma'/I'$ are Koszul algebras then $\Lambda \otimes_K \Lambda'$ is a Koszul algebra.* □

3   Fundamental Results on Koszul Modules

We now turn our attention to modules. We will always assume that all modules are graded Λ-modules and all Λ-homomorphisms are graded homormorphism of degree 0. That is, if $f : M \to N$ then for all $n \in \mathbf{Z}, f(M_n) \subseteq N_n$. We denote the category graded Λ-modules and degree 0 homomorphisms by $\mathrm{Gr}\,\mathrm{mod}(\Lambda)$. We have a functor $\mathcal{E} : \mathrm{Gr}\,\mathrm{mod}(\Lambda) \to \mathrm{Gr}\,\mathrm{mod}(\Lambda^{op})$ given by $\mathcal{E}(M) = \amalg_{n \geq 0} \mathrm{Ext}^n_\Lambda(M, \Lambda/\mathbf{r}) = \mathrm{Ext}^*_\Lambda(M, \Lambda/\mathbf{r})$. In general, even if M is generated in degree 0, the same need not be true for $\mathcal{E}(M)$. This leads to the following definition. We say a graded Λ-module is a *Koszul module* if $\mathcal{E}(M)$ has the property that for each $n \geq 0$, $\mathrm{Ext}^1_\Lambda(\Lambda/\mathbf{r}, \Lambda/\mathbf{r}) \cdot \mathrm{Ext}^n_\Lambda(M, \Lambda/\mathbf{r}) = \mathrm{Ext}^{n+1}_\Lambda(M, \Lambda/\mathbf{r})$. The product is the Yoneda product. In particular, a Koszul module is generated in degree 0 as a graded $E(\Lambda)$-module. If Λ is a Koszul algebra, then the converse is true; that is, M is a Koszul module if and only if M is generated in degree 0. If X is just a Λ-module (not necessarily gradeable) then we say X is a *quasi-Koszul module* if $\mathcal{E}(M)$ has the property that for each $n \geq 0$, $\mathrm{Ext}^1_\Lambda(\Lambda/\mathbf{r}, \Lambda/\mathbf{r}) \cdot \mathrm{Ext}^n_\Lambda(M, \Lambda/\mathbf{r}) = \mathrm{Ext}^{n+1}_\Lambda(M, \Lambda/\mathbf{r})$.

Our first result is a module version of Theorem 2.1.

Theorem 3.1 *[14, Thm 3.3] Let $\Lambda = K\Gamma/I$ be a graded algebra with the length grading and let M be a graded Λ-module generated in degree 0. Then the following statements are equivalent:*

1. M is a Koszul module.

2. M has a linear graded projective resolution.

□

As an immediate consequence, we see that Λ is a Koszul algebra if and only if $\Lambda/\mathbf{r}$ is a Koszul module. The next proposition presents some homological properties of Koszul modules.

Proposition 3.2 [14, Props. 5.2,5.3] Assume that $\Lambda = K\Gamma/I$ is a graded algebra. Let $0 \to A \to B \to C \to 0$ be a short exact sequence of graded Λ-modules. Assume that $\mathbf{r}B \cap A = \mathbf{r}A$. Then if A and C are Koszul modules it follows that B is a Koszul module. If A and B are Koszul modules then so is C. □

Before turning to the category of Koszul modules in a Koszul algebra, we present a final important homological result. The following result is one of the main tools used in proving many of the results of this and the last section. If M is a graded Λ-module, we will denote the nth-graded syzygy of M by $\Omega^n(M)$.

Theorem 3.3 [14, Thm 5.6] Let $\Lambda = K\Gamma/I$ be a Koszul algebra. Then for any finitely generated graded indecomposable, projective Λ-module P and for any positive integer n,

1. for each $t \geq 1$, there exists a short exact sequence

$$0 \to \Omega^i(\mathbf{r}^{n-1}P) \to \Omega^i(\mathbf{r}^{n-1}P/\mathbf{r}^n P) \to \Omega^{i-1}(\mathbf{r}^n P) \to 0,$$

and

2. for each $t \geq 1$, there is a short exact sequence

$$0 \to Ext_\Lambda^{i-1}(\mathbf{r}^n P, \Lambda/\mathbf{r}) \to Ext_\Lambda^i(\mathbf{r}^{n-1}P/\mathbf{r}^n P, \Lambda/\mathbf{r}) \to Ext_\Lambda^i(\mathbf{r}^{n-1}P, \Lambda/\mathbf{r}) \to 0.$$

□

For the remainder of this section we assume that $\Lambda = K\Gamma/I$ is a Koszul algebra. We let $\mathcal{K}_\Lambda$ denote the category of Koszul modules over Λ and degree 0 homomorphisms. Similarly, we let $\mathcal{K}_{E(\Lambda)}$ denote the category of right Koszul modules over $E(\Lambda)$ and degree 0 homomorphisms. We denote the kth-shift of a graded module M by $M(-k)$. That is, $M(-k) = N$ where $N_n = M_{n+k}$. Finally, let $\mathbf{r}_{E(\Lambda)}$ denote the graded Jacobson radical of $E(\Lambda)$. Thus $\mathbf{r}_{E(\Lambda)} = \coprod_{n \geq 1} Ext_\Lambda^n(\Lambda/\mathbf{r}, \Lambda/\mathbf{r})$.

Proposition 3.4 *[15, Prop 5.1] Let Λ be a Koszul algebra. Then $\mathcal{E} : \mathcal{K}_\Lambda \to$ Gr mod$(E(\Lambda)^{op})$ given by $\mathcal{E}(M) = Ext_\Lambda^*(M, \Lambda/\mathbf{r})$ satisfies the following properties:*

1. *If W is a semisimple, graded, finitely generated Λ-module generated in degree 0 then $W \in \mathcal{K}_\Lambda$ and $\mathcal{E}(W)$ is a finitely generated, graded projective $E(\Lambda)^{op}$-module generated in degree 0. Hence $W \in \mathcal{K}_{E(\Lambda)}$.*

2. *If P is a finitely generated, graded projective Λ-module generated in degree 0 then $\mathcal{E}(P)$ is a finitely generated semisimple $E(\Lambda)^{op}$-module generated in degree 0. Hence $\mathcal{E}(P) \in \mathcal{K}_{E(\Lambda)}$.*

3. *If $M \in \mathcal{K}_\Lambda$ then $\mathbf{r}^k M(-k) \in \mathcal{K}_\Lambda$ for $k \geq 1$ and also $\Omega_{E(\Lambda)}\mathcal{E}(M)(-k) = \mathcal{E}(\mathbf{r}^k M(-k))$ for $k \geq 1$.*

4. *If $M \in \mathcal{K}_\Lambda$ then $\mathcal{E}(M) \in \mathcal{K}_{E(\Lambda)}$ and $\mathbf{r}_{E(\Lambda)}\mathcal{E}(M) = \coprod_{n \geq 1} Ext_\Lambda^n(M, \Lambda/\mathbf{r})$.*

5. *If $M \in \mathcal{K}_\Lambda$ then $\mathcal{E}(\Omega(M)(-1)) = \mathbf{r}_{E(\Lambda)}\mathcal{E}(M)(-1)$ and $\mathcal{E}(\Omega^k(M)(-k)) = \mathbf{r}_{E(\Lambda)}\mathcal{E}(M)(-k)$ for $k \geq 1$.*

□

As this proposition hints at, there is a duality here.

Theorem 3.5 *[15, Thm 5.2] Let $\Lambda = K\Gamma/I$ be a Koszul algebra with Yoneda algebra $E(\Lambda)$. Let $\mathcal{K}_\Lambda$ and $\mathcal{K}_{E(\Lambda)}$ denote the categories of Koszul modules in Λ and $E(\Lambda)^{op}$ respectively. Then the contravariant functor $\mathcal{E} : \mathcal{K}_\Lambda \to \mathcal{K}_{E(\Lambda)}$ given by $\mathcal{E}(M) = \coprod_{n \geq 0} Ext_\Lambda^n(M, \Lambda/\mathbf{r})$ is a duality.* □

4   The Nongraded Case

Throughout this section, Λ will denote a semiperfect Noetherian K-algebra with Jacobson radical $\mathbf{r}$. Many of the results about Koszul algebras remain valid in this setting. We begin by defining what a "linear resolution" would mean in this nongraded case. Let M be a finitely generated Λ-module and let

$$(*)   \cdots \to P_n \xrightarrow{f_n} P_{n-1} \to \cdots P_0 \xrightarrow{f_0} M \to 0$$

be a Λ-projective resolution of M with each P_i finitely generated. We say $(*)$ is a *linear resolution of M* if for each $i \geq 0$, $\ker(f_i) \subseteq \mathbf{r}P_i$ and $\mathbf{r}\ker(f_i) = \mathbf{r}^2 P_i \cap \ker(f_i)$. It is not hard to show that in the graded case, minimal graded linear resolutions have these two properties (with $\mathbf{r}$ being the graded Jacobson radical). We can now state our first result.

Theorem 4.1 *[14, Thm 4.4] If Λ is a semiperfect, Noetherian K-algebra and M is a finitely generated Λ-module, then the following statements are equivalent:*

1. *M is a quasi-Koszul module.*

2. *M has a linear resolution.*

□

In this nongraded case, we need another concept. We say a finitely generated Λ-module is *strongly quasi-Koszul* if (*) is a minimal projective resolution of M and satisfies for each $i \geq 0$ and $k \geq 1$

$$\mathbf{r}^k \ker(f_i) = \mathbf{r}^{k+1} P_i \cap \ker(f_i).$$

We say a semiperfect Noetherian ring, Λ, is *strongly quasi-Koszul* if every simple Λ-module is strongly quasi-Koszul. Note that in the graded case, every finitely generated Koszul module is strongly quasi-Koszul [14, Lemma 5.1]. We have the following theorem.

Theorem 4.2 *[14, Thm 6.1] Let Λ be a Noetherian, semiperfect, strongly quasi-Koszul ring. Then the Yoneda algebra $E(\Lambda)$ is a graded quasi-Koszul ring.* □

In [14], there are nongraded versions of Proposition 3.2 and Theorem 3.3 of the previous section of this paper. Further study of the nongraded case is called for. The relationship between a strongly quasi-Koszul semiperfect Noetherian ring Λ and its associated graded ring, $G_{\mathbf{r}}(\Lambda) = \Lambda/\mathbf{r} + \mathbf{r}/\mathbf{r}^2 + \mathbf{r}^2/\mathbf{r}^3 + \cdots$ should be investigated. Assuming that Λ is strongly quasi-Koszul, semiperfect and Noetherian, is the associated graded ring Koszul? In this case, is the associated graded ring $G_{\mathbf{r}}(\Lambda)$ isomorphic to the Yoneda algebra of the Yoneda algebra of $G_{\mathbf{r}}(\Lambda)$?

5   Global Dimensions 1 and 2 and Auslander Algebras

Suppose that $\Lambda = K\Gamma$. Then Λ is global dimension 1 and since $I = (0)$ we see that Λ is a Koszul algebra. Applying Thm 2.4, we see that $E(\Lambda)$ is isomorphic to $K\Gamma^{op}/(L^\circ)^2$, where L° is the ideal in $K\Gamma^{op}$ generated by the arrows of Γ^{op}. Then next result describes the Koszul modules in this case.

Proposition 5.1 *[15, Props 6.1, 6.2] Let $\Lambda = K\Gamma$ and **r** be the ideal generated by the arrows of $K\Gamma$. Then*

1. *A graded Λ-module M generated in degree 0 is a Koszul module if and only $\mathbf{r}M$ is projective.*

2. *If M is a (graded) Koszul Λ-module then every finitely generated submodule of M is quasi-Koszul.*

3. *If M is a graded Λ-module with minimal graded projective resolution $0 \to P_1 \xrightarrow{f} P_0 \to M \to 0$, then the following statements are equivalent:*

   (a) M is a Koszul module

   (b) for every split monomorphism $g : Q \to P_1$ the $\mathrm{coker}(fg)$ is a Koszul module.

   (c) for every indecomposable projective module Q and split monomorphism $g : Q \to P_1$, the $\mathrm{coker}(fg)$ is a Koszul module.

□

Note that, since the global dimension of $K\Gamma/I$ is at least 2, if $I \neq (0)$ and $I \subset J^2$ where J is the ideal generated by the arrows in Γ, the only case of global dimension 1 is the path algebra $K\Gamma$. The next result that shows that every quadratic global dimension 2 algebra is a Koszul algebra.

Theorem 5.2 *[14, Thm 7.2] Let $\Lambda = K\Gamma/I$ be a K-algebra. Assume that I is generated by quadratic elements and that the global dimension of Λ is 2. Then Λ is a Koszul algebra.* □

The Auslander algebras are a class of algebras of global dimension 2. But, in general, they need not be quadratic. They are always quasi-Koszul. Recall the definition of the Auslander algebra. Let A be a finite dimensional algebra over a field K. We say A is of *finite representation type* if there are only a finite number of nonisomorphic indecomposable finitely generated A-modules. Suppose that A is of finite representation type and let $X_1, \ldots, X_n$ be full set of nonisomorphic indecomposable A-modules. Set $X = \coprod_{i=1}^{n} X_i$. Then the *Auslander algebra of A* is $\Lambda = \mathrm{End}_A(X)^{op}$.

The (non-semisimple) Auslander algebra is always of global dimension 2,[2, Prop 5.2]. We have the following result.

Theorem 5.3 *[14, Thm 7.4] Assume that K is an algebraically closed field. Let A be a finite dimensional K-algebra of finite representation type. Then the Auslander algebra of A is a quasi-Koszul algebra.*$_\Box$

We end this section by describing some of the properties of the Yoneda algebra of an Auslander algebra. Keeping the notations above, if A is an algebra of finite representation type and M is an A-module, then $\text{Hom}_A(M, X)$ is a projective Λ-module where $\Lambda = \text{End}_A(X)^{op}$ is the Auslander algebra of A. If M is an indecomposable A-module then $\text{Hom}_A(M, X)$ is an indecomposable projective Λ-module. We let $[M]$ denote the simple Λ-module which $\text{Hom}_A(M, X)$ maps onto.

Theorem 5.4 *[14, Thm 8.2] Let A be a finite dimensional K-algebra of finite type over an algebraically closed field K. Let Λ be the Auslander algebra of A and let $E(\Lambda)$ denote the Yoneda algebra of Λ. Then a projective $E(\Lambda)^{\text{op}}$-module $P_M = \coprod_{k=0}^{2} Ext_\Lambda^k([M], \Lambda/\mathbf{r})$ is injective if and only if the simple Λ-module $[M]$ corresponds to a noninjective A-module M. In particular, $E(\Lambda)$ is Loewy length 3 and each projective of Loewy length 3 is injective. Moreover, $E(\Lambda)$ is 1-Gorenstein.*$_\Box$

A complete description of the Yoneda algebra of an Auslander algebra would be interesting. In particular, suppose that Σ is a Koszul algebra satisfying the following properties:

1. The Loewy length of Σ is 3.

2. Each indecomposable projective Σ-module of Loewy length 3 is injective.

3. Σ is 1-Gorenstein.

What extra conditions imply that $E(\Sigma)$ is an Auslander algebra?

6 Gröbner Bases and Koszul Algebras

In this section we present a sufficient condition for an algebra to be a Koszul algebra. We briefly introduce Gröbner bases in path algebras. For more details, see [9]. We need an admissible order on the paths. Let Γ be a finite directed graph and let B denote the set of finite directed paths in Γ. Recall that we view the vertices as paths of length 0. A well ordering $<$ on B is called an *admissble order* if it satisfies the following properties:

C1 if $p, q \in B$ and $p < q$ then $pr < qr$ for all $r \in B$ such that both pr and qr are nonzero.

C2 if $p, q \in B$ and $p < q$ then $sp < sq$ for all $s \in B$ such that both sp and sq are nonzero.

C3 if $p, q, r, s \in B$ and $p = rqs$ then $q \leq p$.

There are many admissible orders possible. For example, if we arbitraryly order the vertices of Γ, and arbitrarily order the arrows then we define a *length-lexicographic* order as follows. If both p and q are both vertices or both arrows, use the above arbitrary order. If $p, q \in B$ then $p < q$ if either the length of p is less than the length of q or $p = a_n \ldots a_1$ and $q = b_n \ldots b_1$ with $a_i, b_i \in \Gamma_1$ and for some $i_0 \geq 1$, $a_j = b_j$ if $j < i_0$ and $a_{i_0} < b_{i_0}$. It is easy to check that $<$ is an admissible order.

We will assume for the remainder that $<$ is an admissible ordering on B. Let $x = \sum_i \alpha_i p_i \in K\Gamma - \{0\}$ where $\alpha_i \in K - \{0\}$ and the p_i are distinct paths. We let $\text{tip}(x) = p_i$ if $p_i \geq p_j$ for all paths p_j in x. If p and q are paths, we say p *divides* q and write $p \mid q$ if there exist paths r and s such that $q = rps$.

Let I be an ideal in $K\Gamma$ and suppose that $\mathcal{G} = \{f_1, \ldots, f_n\}$ generate I. We say that $\mathcal{G}$ is a *Gröbner basis for I (with respect to $<$)* if for each nonzero $f \in I$ there exists an $f_i \in \mathcal{G}$ such that $\text{tip}(f_i) \mid \text{tip}(f)$. There is an easy check to determine if a finite set of elements of I is a Gröbner basis. In general, given an ideal I, there is no finite Gröbner basis; but, if $K\Gamma/I$ is finite dimensional, then there is a finite Gröbner basis and a finite algorithm to find this basis, see [9, 20]. Furthermore, if there is a finite Gröbner basis then there is a finite algorithm to find the basis. If there is a Gröbner basis consisting of quadratic elements, then there is a finite Gröbner basis consisting of quadratic elements and hence can be found algorithmically. The next result has many interesting consequences, one of which will be given in the next section.

Theorem 6.1 *[13] Let $\Lambda = K\Gamma/I$ be a graded algebra with the length grading and let $<$ be an admissible order on the paths of Γ. Suppose that I has a Gröbner basis consisting of quadratic elements. Then Λ is a Koszul algebra.*

□

The actual result in [13] shows that the Anick-Green projective resolution of $\Lambda/\mathbf{r}$ (see [1]) is a minimal projective graded resolution. Thus, in case there is a quadratic Gröbner basis, the linear resolution of $\Lambda/\mathbf{r}$ can be constructed explicitly. In [13], the theorem is used to show algebras of maximal minors and algebras obtained from straightening rules are Koszul algebras. We also remark that the above theorem can also be used to show that quadratic algebras with Poincaré-Birkhoff-Witt bases are Koszul algebras ([12]).

7   Generalized Preprojective Algebras

In this section we present a new class of Koszul algebras. We begin with an undirected graph G. We allow G to have multiple edges and loops. Give G some orientation which we denote by $\vec{G}$. Let $a_1, \ldots, a_m$ be the arrows in $\vec{G}$. For each arrow $a_i : v \to w$, let $a_i^o : w \to v$ be a new arrow in the opposite direction as a_i. We let Γ_G be the directed graph with vertex set the same as G and with arrow set $\{a_1, \ldots, a_m\} \cup \{a_1^o, \ldots, a_m^o\}$. Note that Γ_G is independent of the orientation chosen for G. We now define a set of elements in $K\Gamma_G$. For each vertex $v \in (\Gamma_G)_0$, let $r_v = \sum_{i=0}^m \sigma_i(v) a_i a_i^o + \tau_i(v) a_i^o a_i$ where $\sigma_i(v)$ and $\tau_i(v)$ are elements in K. We require $\sigma_i(v) \neq 0$ if and only if the origin of a_i^o is v and $\tau_i(v) \neq 0$ if and only if the origin of a_i is v. Let I_G be the ideal in $K\Gamma_G$ generated by the $\{r_v\}_{v \in (\Gamma_G)_0}$. We call $\Lambda = K\Gamma_G / I_G$ a *generalized preprojective algebra*. If the nonzero σs and τs are all 1 then Λ is called a *preprojective algebra*. The preprojective algebras have appeared in the study of the representation theory of finite dimensional algebras, [6, 10], and more recently, in some geometrical considerations of Nakajima [21].

We give an example for clarification. Suppose that G is the graph

$$
\begin{array}{ccccc}
v_1 & - & v_2 & & \\
| & & | & & \\
v_3 & - & v_4 & - & v_5
\end{array}
$$

Now give G some orientation. For example, $\vec{G}$ might be

$$
\begin{array}{ccccc}
v_1 & \xrightarrow{a_1} & v_2 & & \\
\downarrow a_2 & & \downarrow a_3 & & \\
v_3 & \xrightarrow{a_4} & v_4 & \xrightarrow{a_5} & v_5
\end{array}
$$

Then Γ_G will be the graph

$$
\begin{array}{ccccc}
& a_1 & & & \\
v_1 & \xrightarrow{\ \ \ } & v_2 & & \\
\bullet & \xleftarrow{\ \ \ } & & & \\
& a_1^o & & & \\
a_2^o \updownarrow a_2 & & a_3^o \updownarrow a_3 & & \\
& a_4 & & a_5 & \\
v_3 & \xrightarrow{\ \ \ } & v_4 & \xrightarrow{\ \ \ } & v_5 \\
\bullet & \xleftarrow{\ \ \ } & & \xleftarrow{\ \ \ } & \bullet \\
& a_4^o & & a_5^o &
\end{array}
$$

The generators of the ideal I_G are

$$
\begin{aligned}
r_{v_1} &= \tau_1(v_1)a_1^o a_1 + \tau_2(v_1)a_2^o a_2 \\
r_{v_2} &= \sigma_1(v_2)a_1 a_1^o + \tau_3(v_2)a_3^o a_3 \\
r_{v_3} &= \sigma_2(v_3)a_2 a_2^o + \tau_4(v_3)a_4^o a_4 \\
r_{v_4} &= \sigma_3(v_4)a_3 a_3^o + \sigma_4(v_4)a_4 a_4^o + \tau_5 a_5^o a_5 \\
r_{v_5} &= \sigma_5 a_5 a_5^o
\end{aligned}
$$

where the σs and the τs are some fixed nonzero elements of K.

We will show that if G is a graph with no connected component a tree, then each generalized preprojective algebra $K\Gamma_G / I_G$ is a Koszul algebra. We will find a specific length-lexicographic order for the paths of Γ_G. For this we need some preliminary results. Recall that the *degree of a vertex* v is twice the number of loops at v plus the number of edges that are not loops of which v is an endpoint.

Proposition 7.1 *Let G be an undirected finite graph and let G_0 and G_1 denote the vertices and edges respectively of the graph G. No connected component of G is a tree if and only if there is a set function $\phi : G_0 \to G_1$ such that ϕ is one-to-one and for each vertex $v \in G_0$, v is an endpoint of the edge $\phi(v)$.*

Proof. Suppose some connected component, G', of G is a tree. Then the number of vertices of G' is one more than the number of edges in G'. Hence no such ϕ can exist.

Now suppose that no connected component of G is a tree. We proceed by induction on the number of vertices in G. If $|G_0| = 1$ then there must be a loop at the vertex of G and the result follows. Now assume the result for graphs with $n - 1$ vertices, none of whose connected components is a tree. Assume that $|G_0| = n$.

First suppose that G has a vertex of degree 1. That is, there is a vertex, say v, which is an endpoint of exactly one edge, say e. Set $\phi(v) = e$. Remove

v and e from G to obtain a new graph G^*. It is clear that no component of G^* is tree since the degree of v is 1. The existence of ϕ follows by induction.

Next, suppose that G has no vertex of degree 1. Assume, without loss of generality, that G is connected. Choose a vertex $v \in G_0$ and an edge $e \in G_1$ such that v is an endpoint of e. Let G^* be the graph obtained from G after removing v and all the edges with at least one endpoint being v. Let $\phi(v) = e$. If no component of G^* is a tree, we are done by induction. Hence, assume that G^* has a component which is a tree. Let G^{**} be a connected component of G^* which is a tree. Then G^{**} has at least 2 vertices of degree 1. Both of these cannot be endpoints of e. Hence there is a vertex, say v^*, of G^{**} of degree 1 in G^{**} such that v^* is not an endpoint of e.

We orient the edges in G^{**} as follows. If e' is an edge in G^{**} with endpoints v_1 and v_2, orient e' from v_1 to v_2 if v_1 is closer to v^* than v_2, otherwise, orient e' from v_2 to v_1. This makes sense, since G^{**} is a tree and there is an edge between v_1 and v_2. For each vertex $w \in G^{**} - \{v^*\}$ choose an edge $e_w \in G^{**}$ such that the terminus of e_w is w under the above construction. Now define $\phi(w) = e_w$. We have defined ϕ for all vertices of G^{**} except for v^*. But, in G, the degree of v^* is at least 2, where as in G^{**} it is degree 1. Thus, there is an edge e^* with endpoints v and v^*, which, by construction, is not e. Thus define $\phi(v^*) = e^*$ and we are done. $\square$

We can now state the main result of this section.

Theorem 7.2 *Let K be a field, G be an undirected graph such that no connected component of G is a tree. Let $\Lambda = K\Gamma_G/I_G$ be a generalized pre-projective algebra whose construction is described above. Then there is a length-lexicographic order on the paths in Γ_G such that there is a quadratic Gröbner basis for I_G. Hence Λ is a Koszul algebra.*

Proof. Let $\phi : G_0 \to G_1$ be a one-to-one set map so that for each vertex $v \in G_0$, v is an endpoint of $\phi(v)$; which exists by Proposition 7.1. Let Γ_G be the quiver obtained as above after we orient G. Let $\{a_1, \ldots, a_m\}$ be the (oriented) arrows of G (under some fixed orientation) and so the arrows of Γ_G are $\{a_1, \ldots, a_m\} \cup \{a_1^o, \ldots, a_m^o\}$ where if $a_i : v \to w$ then $a_i^o : w \to v$. Finally, for $i = 1, \ldots, m$ let e_i be the edge in G associated to a_i.

We now define the length-lexicographic order we will use. For this we need to order the vertices and the arrows of Γ_G. Order the vertices arbitrarily. Now order the arrows as follows. Given a vertex v, consider $\phi(v) = e$. There are two arrows associated to e of the form a and a^o. Only one of these has terminus v unless e is a loop. Call this arrow a_v if e is not a loop and choose a_i as a_v if e is a loop. Order the arrows of Γ_G so that the arrows of the form

a_v are larger than any arrow that is not of the form a_v. Now consider the generators $\{r_v\}$ of I_v. We claim, under an order described above, $\{r_v\}$ is a quadratic Gröbner basis. If so, Λ is a Koszul algebra by Theorem 6.1. To show this we need only show that there are no overlaps of tips of the $\{r_v\}$ (see [9]).

By construction, only one edge, e, with endpoint v is of the form $\phi(v)$. So there is exactly one arrow labelled a_v with terminus v. Thus, the other arrows with terminus v are smaller than a_v. Hence is $r_v = ca_v b + \sum \alpha_i p_i$ where $b \in (\Gamma_G)_1$, $c, \alpha_i \in K^*$ and p_i are paths of length 2 of the form $a_i a_i^o$ or $a_i^o a_i$ with no arrow labelled a_v. Note that b is not an a_w either since it is also associated to e by the terminus not v (or e is a loop and $b = a_v^o$). It follows that $\operatorname{tip}(r_v) = a_v b$ and there are no overlaps. $\square$

We note that in case G is a tree, R. Martínez-Villa has announce results which classify in which cases the preprojective algebra is Koszul. He shows that the graph G must not be a Dynkin diagram and his techniques are very different than those of this paper.

We end this section with an example that shows that there are Koszul algebras such that there is no order that yields a quadratic Gröbner basis but that the Yoneda algebra has a quadratic Gröbner basis and we get a quick proof that the original algebra is Koszul.

Let K be a field, not of characteristic 2 and let $K\Gamma/I$ be the algebra $K<x,y>/I$ where I is the ideal generated by $x^2 - y^2$ in the free associative algebra in two noncommuting variables x and y, $K<x,y>$. Note the Γ is the graph with one vertex and two loops. Under no admissible ordering is the Gröbner basis of I quadratic. But note that if I_2 is the span of $x^2 - y^2$ in the vector space of generated by paths of length 2, then $I_2^\perp$ is generated by $xy, yx, x^2 + y^2$. It is easy to show that $\{xy, yx, x^2 + y^2\}$ is in fact the reduced Gröbner basis for the ideal I^* generated by $I_2^\perp$. Hence, $K<x,y>/I^*$ is a Koszul algebra by Theorem 6.1. Now by Theorems 2.2 and 2.4, the Yoneda algebra $E(K<x,y>/I^*)$ is a Koszul algebra which is isomorphic to $K<x,y>/<(I^*)_2^\perp>$. But $(I^*)_2^\perp = (I_2^\perp)^\perp = I_2$. Thus $\Lambda = E(K<x,y>/I^*)$ is a Koszul algebra.

References

[1] D. Anick and E.L. Green, *On the homology of path algebras*, Comm. in Algebra, **15**, (1985), 641-659.

[2] M. Auslander, I. Reiten, and S. Smalø, *Representation Theory of Artin*

Algebras , Canbridge Studies in Advanced Math., **36**, (1995), Cambridge Univ. Press.

[3] L. Avramov, and J. Herzog, *The Koszul algebra of a codimension* 2 *embedding*, Math. Z. **175** (1980), 249-260.

[4] J. Backelin, R. Froberg, *Koszul algebras, Veronese subrings and rings with linear resolutions*, Rev. Roumaine Math. Pures Appl. **30** (1980), 85-97.

[5] A.I. Bondal, *Helices, representations of quivers and Koszul algebras*, London Math. Soc. Lecture Note Ser. **148**, (1990), 75-95.

[6] D. Baer, *Noetherian Categories and representation theory of hereditary Artin algebras*, Comm in Algebra, **15**, (1985) 247-258.

[7] E. Cline, B. Parshall, and L. Scott, *Finite dimensional algebras and highest weight categories*, J. Reine Angew. Math. **391** (1988), 85-99.

[8] A. Beilinson, V. Ginsburg, & W. Soergel, *Koszul Duality Patterns in Representation Theory*, preprint.

[9] D.R. Farkas, C. Feustel, and E.L. Green, *Synergy in the theories of Gröbner bases and path algebras*, Canad. J. of Mathematics, **45** (1993), 727-739.

[10] I.M. Gelfand, and V.A. Ponomarev, *Model algebras and representations of graphs*, Funct. Anal. Appl. **13**, (1979), 157-166.

[11] E.L. Green, Representation theory of tensor algebras, J. Algebra, **34**, (1975), 136-171.

[12] E.L. Green, Poincaré-Birkhoff-Witt bases and Gröbner bases, preprint.

[13] E.L. Green and R. Huang, *Projective resolutions of straightening closed algebras generated by minors*, Adv. in Math. **110** (1995), 314-333.

[14] E.L. Green, and R. Martínez Villa, *Koszul and Yoneda algebras*, Canad. Math. Soc., Conference Proceedings **18** Representation Theory of Algebras, (1994), 247-298.

[15] E.L. Green, and Martínez Villa, *Koszul and Yoneda algebras II*, to appear.

[16] R. Hartshone: *Residues and Duality*, LNM 20, Springer-Verlag, (1966).

[17] C. Löfwall, *On the subalgebra generated by the one-dimensional elements in the Yoneda ext-algebra*, LNM 1183, Springer-Verlag, (1986), 291-338.

[18] R. Martínez Villa, *Applications of Koszul algebras: the preprojective algebra*, Canad. Math. Soc., Conference Proceedings **18** Representation Theory of Algebras, (1994), 487-504.

[19] S. McLane, *Homology*, Springer-Verlag, 1963.

[20] F. Mora, *Gröbner bases for non-commutative polynomial rings*, Proc. AAECC3 L.N.C.S. **229** (1986).

[21] H. Nakajima, *Varieties associated with quivers*, to appear Canad. Math. Soc. Proceedings of ICRA VII, Mexico, 1994.

[22] S. Priddy, *Koszul resolutions*, Trans. AMS, **152** (1970), 39-60.

[23] M. Rosso, *Koszul resolutions and quantum groups*, Nuclear Phys. B Proc. Suppl., **18b** (1990), 269-276.

[24] A. Vishik, and M. Finkelberg, *The coordinate ring of general curves of genus $g \geq 5$ iss Koszul*, J. Algebra, **162**, (1993), 535-539.

[25] Y. Yoshino, *Modules with linear resolutions over a polynomial ring in two variables*, Nagoya Math. J. **113**, (1989), 89-98.

Old and recent work with Maurice

Idun Reiten

I met Maurice in 1970 in Urbana, Illinois, where I was a graduate student of Robert Fossum. Maurice came to Urbana to work on representation theory. He gave an advanced course, where he usually lectured, with great enthusiasm, on what he had proved since the previous session. I saw something like this for the first time, and I found the mathematics, and Maurice's approach to it, very exciting. I had a fellowship from the Norwegian Research Council, which also could support me beyond my Ph.D. in 1971, and I wanted to go to Brandeis to work with Maurice. When I finally had the courage to ask him if this would be possible, he said that he had been thinking about the same thing, but he was not interested if I did not have the courage and initiative to ask. This was the start of a collaboration which lasted until Maurice's death, and which produced about 40 papers, some with additional coauthors. Maurice viewed the semester in Urbana as the start of his work on representation theory, as indicated by what he wrote to me:

(March 22, 74) *It was really nice being back in Urbana, the place where all of this mathematics started.*

I shall here give a survey of some of the main parts of our joint work, in an expanded version of my lecture at the Maurice Auslander Memorial Conference. The organization is according to topics, usually in the order in which the work started. No attempt is made to review the contributions of others on these topics, and the list of references is restricted to cover my work with Maurice. In connection with almost split sequences and irreducible maps I will give some citations from letters to show the way Maurice viewed our work and how he was concerned about the reactions of others.

In most of the topics there are strong ties to Maurice's earlier work. Throughout he kept his interest in homological algebra, including functorial approaches. He was always interested in impact on commutative ring theory, or more generally higher dimensional noncommutative ring theory, even though the representation theory of artin algebras is the main theme during this period.

Mostly Λ will be an artin algebra over a commutative artin ring R, for example a finite dimensional algebra over a field k, and $\mathrm{mod}\,\Lambda$ will denote the category of finitely generated (left) Λ-modules. Relevant examples to keep in mind are group algebras kG where G is a finite group, finite dimensional factor algebras of the polynomial ring $k[X_1, \ldots , X_n]$ and matrix algebras like $\begin{pmatrix} k & 0 & 0 \\ k & k & 0 \\ k & 0 & k \end{pmatrix}$. More generally we have factor algebras of path algebras of quivers,

for example $\Lambda = k\Gamma/I$ where Γ is the quiver $\cdot \overset{\alpha}{\to} \cdot \overset{\beta}{\underset{\gamma}{\to}} \cdot$ and I is the ideal

generated by $\beta\alpha$. Denote by D the ordinary duality between $\mathrm{mod}\,\Lambda$ and $\mathrm{mod}\,\Lambda^{\mathrm{op}}$, which for a k-algebra is $\mathrm{Hom}_k(\ , k)$, and by Tr the transpose which we define below.

1   Stable equivalence

After spending the fall of 1970 in Urbana, Maurice went to Europe for the spring of 1971. Especially his conversations with J. A. Green and P. Gabriel were to influence our work when we met again at Brandeis in the fall of 1971 and our real collaboration began.

Maurice had been interested in the module category modulo projectives $\underline{\mathrm{mod}}\Lambda$. This category is the natural domain of definition for the functors $\mathrm{Ext}^i_\Lambda(\ , C)$ where C is in $\mathrm{mod}\,\Lambda$ and $i > 0$, and for his favourite functor, the transpose Tr. Recall that if $P_1 \to P_0 \to C \to 0$ is a projective presentation of C in $\mathrm{mod}\,\Lambda$ (for a noetherian ring Λ), then $\mathrm{Tr}\,C$ is defined up to projective summands by the exact sequence $0 \to C^* \to P_0^* \to P_1^* \to \mathrm{Tr}\,C \to 0$, where $X^* = \mathrm{Hom}_\Lambda(X, \Lambda)$. For artin algebras we usually define $\mathrm{Tr}\,C$ by using a minimal projective presentation for C. Then Tr does not in general give rise to a functor from $\mathrm{mod}\,\Lambda$ to $\mathrm{mod}\,\Lambda^{\mathrm{op}}$, but there is a duality $\mathrm{Tr} : \underline{\mathrm{mod}}\Lambda \to \underline{\mathrm{mod}}\Lambda^{\mathrm{op}}$. Maurice had first used the transpose many years earlier in the context of regular local rings and also for noncommutative noetherian rings in work with Bridger.

Two artin algebras Λ and Λ' are *stably equivalent* if the categories modulo projectives $\underline{\mathrm{mod}}\Lambda$ and $\underline{\mathrm{mod}}\Lambda'$ are equivalent. The algebras Λ and Λ' may be very different from a homological point of view even if they are stably equivalent, but it is easy to see that Λ is of finite representation type if and only if Λ' is.

From J. A. Green Maurice had learned about interesting instances of stable equivalences occuring for group algebras, induced by Green correspondence. On the other hand Gabriel had told Maurice about his classification of the quivers where the category of representations has only a finite number of indecomposable objects up to isomorphism, in terms of Dynkin diagrams. For an algebraically closed field k this took care of the classification of the hereditary algebras and radical squared zero k-algebras of finite representation type, that is, having only a finite number of indecomposable modules up to isomorphism. When $\underline{r}^2 = 0$ for Λ, where $\underline{r}$ denotes the radical, there is an associated quiver, called the separated quiver of Λ, which is a disjoint union of Dynkin quivers if and only if Λ is of finite representation type. For example if $\Lambda = \left\{ \left(\begin{smallmatrix} \alpha & 0 & 0 \\ a & b & 0 \\ c & 0 & \alpha \end{smallmatrix} \right) ; \alpha, a, b, c, d \text{ in } k \right\}$, the ordinary quiver is $\bigcirc\!\!\!\!\!\!\!\overset{\bullet\; \to\; \bullet}{\uparrow}$ and the

separated quiver is $\overset{\bullet}{\underset{\bullet\quad\bullet}{\downarrow\searrow}}$, where the path algebra also has radical square

zero (see [1]). We discovered that this gave other interesting examples of stable equivalences. In particular we were led to the following [2].

Proposition 1.1 *An artin algebra Λ with $\underline{r}^2 = 0$ is stably equivalent to the hereditary algebra $\Lambda' = \left(\begin{smallmatrix} \Lambda/\underline{r} & 0 \\ \underline{r} & \Lambda/\underline{r} \end{smallmatrix} \right)$.*

For example $\Lambda = k[X, Y]/(X, Y)^2$ and $\Lambda' = \left(\begin{smallmatrix} k & 0 \\ k \oplus k & k \end{smallmatrix} \right)$ are stably equivalent. In particular this shows that one algebra can have global dimension one and the other infinite global dimension when the algebras are stably equivalent, in strong contrast to Morita equivalence.

More generally, the following was our first joint main result [2].

Theorem 1.2 *An artin algebra Λ is stably equivalent to a hereditary artin algebra if and only if Λ satisfies the following conditions:*

(i) *Each indecomposable submodule of a projective module is simple or projective.*

(ii) *If S is a simple nonprojective submodule of a projective module, then S is a factor of an injective module.*

The method we used in our work on stable equivalence, and in particular for proving this theorem, was functor categories. One drawback about the additive category $\text{mod}\Lambda$ is that it is not an abelian category. If Λ is of finite representation type and M is the direct sum of one copy of each indecomposable object in $\text{mod}\,\Lambda$ up to isomorphism, we consider the algebra $\Gamma = \underline{\text{End}}_\Lambda(M)^{\text{op}}$. Then $\underline{\text{mod}}\Lambda$ is equivalent to the category of projective objects in $\text{mod}\,\Gamma$, which is not only an abelian category, but in this case even a module category for an artin algebra. When Λ is not of finite representation type, this point of view leads to the investigation of functor categories. Here we used and developed further Maurice's earlier work on finitely presented functors. In general the stable category $\underline{\text{mod}}\Lambda$ is equivalent to the projective objects in the category $\mathcal{C}$ of finitely presented contravariant functors from $\text{mod}\,\Lambda$ to abelian groups which vanish on projective objects. This means that the functors F in $\mathcal{C}$ are associated with short exact sequences $0 \to A \overset{f}{\to} B \overset{g}{\to} C \to 0$ giving rise to an exact sequence of functors $0 \to (\,, A) \overset{(\,,f)}{\to} (\,, B) \overset{(\,,g)}{\to} (\,, C) \to F \to 0$ with

F in $\mathcal{C}$. An important feature is that $\mathcal{C}$ has enough projective and injective objects and that both have a nice description, as $\underline{\mathrm{Hom}}(\ ,C)$ and $\mathrm{Ext}^1_\Lambda(\ ,C)$ respectively, for C in $\mathrm{mod}\,\Lambda$. We gave an explicit description of minimal projective presentations and minimal injective copresentations, which made it possible to describe when functors in $\mathcal{C}$ have low projective dimension. For example $\mathrm{pd}\,\mathrm{Ext}^1_\Lambda(\ ,C) \leq 1$ if and only if C is a submodule of a projective module. Such results provide information on stable equivalence since when Λ and Λ' are stably equivalent, the associated functor categories $\mathcal{C}$ and $\mathcal{C}'$ are equivalent categories. Hence projective dimension is preserved. As a consequence we could obtain results used for example to prove Theorem 1.2. As an illustration we cite the following [2].

Proposition 1.3 *Let Λ and Λ' be stably equivalent artin algebras.*

(a) *Λ has no nonzero projective injective modules if and only if Λ' has no nonzero projective injective modules.*

(b) *Λ is 1-Gorenstein (i.e. the injective envelope of Λ is projective) if and only if Λ' is 1-Gorenstein.*

I presented these results from our joint work at the Ohio Conference celebrating the 60th birthday of Zassenhaus in the spring of 1972. Our methods and results were later exploited further by Martinez. For example a central problem, pushed amongst others by Alperin, was the following: Do stably equivalent artin algebras have the same number of nonprojective simple modules up to isomorphism? The main interest was for selfinjective algebras, in particular group algebras. Maurice and I proved it in the "opposite" situation, where one of the algebras has no nonzero projective injective modules and also for algebras stably equivalent to hereditary algebras [2]. Later Martinez has shown that it is sufficient to treat the selfinjective algebras, and he has proved the conjecture for finite representation type.

We kept throughout our collaboration an interest in stable equivalence, and worked occasionally on the above conjecture, without success. We also got back to theoretical work on stable equivalence in connection with almost split sequences and irreducible maps [15, 16]. For example we proved the following, which later turned out to be of interest in connection with triangulated categories [16].

Proposition 1.4 *Let $F : \underline{\mathrm{mod}}\Lambda \to \underline{\mathrm{mod}}\Lambda'$ be an equivalence where Λ and Λ' are selfinjective algebras with no ring summand with radical square zero. Then F commutes with the first syzygy operator Ω^1.*

We also returned to questions about stable equivalence in our investigation of uniserial functors [20]. In particular we studied algebras stably equivalent to Nakayama algebras and biserial algebras.

2 Almost split sequences

During my stay at Brandeis 1971–73 we also started our work on almost split sequences, including proving existence and uniqueness of such sequences for artin algebras. I here give the basic definitions and discuss some of the main points. Also I include several citations from letters written by Maurice in order to give some idea about how he viewed the work, and how he thought others reacted to it.

For an artin algebra Λ an exact sequence $0 \to A \xrightarrow{f} B \xrightarrow{g} C \to 0$ in mod Λ is said to be *almost split* if

(i) it does not split.

(ii) A and C are indecomposable and

(iii) given $h : X \to C$ where X is indecomposable and h is not an isomorphism, then there is some $t : X \to B$ such that $gt = h$.

Our main basic result was the following [8].

Theorem 2.1 *Let Λ be an artin algebra.*

(a) *Let C be an indecomposable nonprojective object in* mod Λ. *Then there is an almost split sequence* $0 \to A \to B \to C \to 0$, *which is unique up to isomorphism, and we have $A \simeq D \operatorname{Tr} C$.*

(b) *Let A be an indecomposable noninjective object in* mod Λ. *Then there is an almost split sequence* $0 \to A \to B \to C \to 0$, *which is unique up to isomorphism, and we have $C \simeq \operatorname{Tr} DA$.*

Like for stable equivalence, the inspiration came from the functorial point of view. The question whether an almost split sequence exists for a given indecomposable module C (without considering the relationship between the end terms) essentially expresses, as is easily seen, that the simple contravariant functors F from mod Λ to abelian groups are finitely presented, that is there is an exact sequence of functors $(\ ,B) \to (\ ,C) \to F \to 0$ with B and C in mod Λ. When Λ is of finite representation type, this is obvious since the category of finitely presented contravariant functors from mod Λ to abelian groups is then just mod Γ for an artin algebra Γ. So the starting point was here whether this property of simple functors, obviously true for finite representation type, holds for artin algebras in general. Part of Maurice's general philosophy and the reason for his interest in finite representation type was that he viewed the class as a test case for what might be true in general.

We had two different approaches to proving the existence of almost split sequences for artin algebras. One was in line with the motivation discussed above, which we formulated more generally in the context of what we called dualizing R-varieties over a commutative artin ring R. Recall that a skeletally small additive R-category $\mathcal{C}$ where idempotents split and each $\operatorname{Hom}(C, C')$ is a finitely generated R-module is a *dualizing R-variety* if there is a duality between the category of finitely presented contravariant and finitely presented covariant functors from C to mod R. This notion was motivated by the important role played by functor categories in the study of artin algebras and the fact that we could prove that the ordinary duality $D : \text{mod}\,\Lambda \to \text{mod}\,\Lambda^{\text{op}}$ extends to a duality between the finitely presented contravariant and covariant functors from mod Λ to mod R. Examples of dualizing R-varieties are mod Λ and the category of finitely generated projective Λ-modules, for an artin algebra Λ, and categories appearing in covering theory for finite dimensional algebras. A way of constructing new examples is based upon the following basic result [3].

Theorem 2.2 *If C is a dualizing R-variety, then the category* $\text{mod}\,C$ *of finitely presented contravariant functors from C to mod R is again a dualizing R-variety.*

In the series of papers [4, 5, 6, 7] we generalized results from artin algebras to dualizing R-varieties, and studied stable equivalence in this context. In particular we extended the results on algebras stably equivalent to hereditary algebras. Our functorial proof of the existence of almost split sequences was done more generally in the context of mod C when C is a dualizing R-variety [3]. Maurice and I also returned recently to work involving dualizing R-varieties, to be discussed in the next section.

The other proof for the existence of almost split sequences was based on a guess of what the relationship should be between the end terms of an almost split sequence, assuming almost split sequences existed. A good test case was the class of group algebras of finite representation type, where the structure of the indecomposable modules was known by work of Kupisch and Janusz. We observed that if $0 \to A \to B \to C \to 0$ is almost split, then A is isomorphic to $\Omega^2 C$, the second syzygy module of C. This could clearly not be the answer in general, since $\Omega^2 C$ need not be indecomposable. On the other hand computations for $k[X,Y]/(X,Y)^2$ with k a field indicated that the answer should be $A = D\,\text{Tr}\,C$, where D is the ordinary duality and Tr the transpose. And indeed, we were able to prove the following, which also gives a useful simple way for testing whether a given exact sequence is almost split [8].

Theorem 2.3 *Let C be an indecomposable nonprojective object in* $\text{mod}\,\Lambda$ *for an artin algebra Λ.*

    (a) $\text{Ext}^1_\Lambda(C, D\,\text{Tr}\,C)$ *is an indecomposable injective* $\underline{\text{End}}(C)$-*module both as a right and left module, and any nonzero element of the simple socle represents an almost split sequence.*

(b) *Let* $0 \to D \operatorname{Tr} C \xrightarrow{f} B \xrightarrow{g} C \to 0$ *be a nonsplit exact sequence in* mod Λ. *Then the sequence is almost split if for each map* $h : C \to C$ *which is not an isomorphism there is some* $s : C \to B$ *with* $gs = h$.

Underlying the existence proof in [8] is the following formula, which has analogues in other contexts where a corresponding formula has been used to prove existence of almost split sequences.

Theorem 2.4 *Let* A *and* C *be in* mod Λ *for an artin algebra* Λ. *Then we have an isomorphism* $D \operatorname{Ext}^1(C, D \operatorname{Tr} A) \simeq \underline{\operatorname{Hom}}(A, C)$ *which is functorial in both variables.*

Maurice and I spent the fall of 1973 at MIT, and during this time we finished our series of papers on dualizing R-varieties in the Advances in Mathematics. This writing process was at the same time part of a therapy for Maurice, who was having problems with his memory after an almost fatal accident in Argentina during the preceeding summer. Maurice had already written two papers in the Communications in Algebra, containing amongst other things his proof of Brauer-Thrall 1. Our paper with the second approach to almost split sequences became part III, and we finished this paper right before the International Conference on Representations of algebras in Ottawa 1974.

While writing up the basic work on almost split sequences we started investigating the maps $f : A \to B$ and $g : B \to C$ occuring in an almost split sequence $0 \to A \xrightarrow{f} B \xrightarrow{g} C \to 0$. This gave rise to what we called irreducible maps. A map $h : X \to Y$ is *irreducible* if h is neither a split monomorphism nor a split epimorphism, and given a commutative diagram

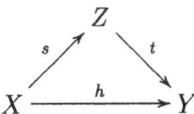

with $h = ts$, then s is a split monomorphism or t is a split epimorphism. The basic connection with almost split sequences is the following [14].

Theorem 2.5 *Let* C *be an indecomposable nonprojective module over an artin algebra* Λ. *Then a map* $h' : B' \to C$ *is irreducible if and only if there is a nonzero map* $h'' : B'' \to C$ *such that there is an almost split sequence* $0 \to A \to B' \amalg B'' \xrightarrow{(h', h'')} C \to 0$.

Our basic work on irreducible maps, and their relationship to almost split sequences, was dealt with in another three papers in the Communications in

Algebra [14, 15, 16]. The writing process was long, due to new developments, changes in points of view and attempts of improvements in the presentation. These papers did not contain very many applications outside the topics of almost split sequences and irreducible maps. But some of the internal computations we made turned out to be useful for applications of the theory. As examples we cite the following [8].

Proposition 2.6 $0 \to A \to B \to C \to 0$ *be an almost split sequence over an artin algebra Λ.*

(a) *B is projective if and only if A is a simple submodule of a projective module which is not a composition factor of $\underline{r}I/\operatorname{soc} I$ for any injective module I.*

(b) *B has an indecomposable projective injective summand P if and only if the almost split sequence is isomorphic to the sequence $0 \to \underline{r}P \to P \amalg \underline{r}P/\operatorname{soc} P \to P/\operatorname{soc} P \to 0$.*

Already while we were working on the basic theory, Maurice applied our results to give an alternative approach to the representation theory of hereditary algebras, in joint work with Platzeck. Also our theory was suited for proving existence of an infinite number of indecomposable modules with a certain property, by constructing chains of irreducible maps. For example, we proved the following [17].

Proposition 2.7 *Let Λ be an indecomposable weakly symmetric artin algebra (that is, $P/\underline{r}P \overset{\sim}{\to} \operatorname{soc} P$ for each indecomposable projective Λ-module P) of infinite representation type.*

If there is one indecomposable Ω-periodic module, then there is an infinite number up to isomorphism.

There were also applications to stable equivalence, and one was already mentioned in the previous section. Also stable equivalences could be used to construct new almost split sequences from old ones [15].

We also mention the following curious property of irreducible maps [14].

Proposition 2.8 *Let $0 \to A \overset{f}{\to} B \overset{g}{\to} C \to 0$ be a nonsplit exact sequence in* $\operatorname{mod} \Lambda$. *Then $f : A \to B$ is irreducible if and only if the map $g : B \to C$ has the property that given any map $h : X \to C$ there is either a map $t : X \to B$ such that $gt = h$ or a map $s : B \to X$ such that $hs = g$.*

On the level of contravariant functors from $\operatorname{mod} \Lambda$ to abelian groups this expresses that for the representable functor $(\ ,C)$ the subfunctor $F = \operatorname{Im}(\ ,g)$ has the property that for any subfunctor $H = (\ ,C)$ we have either $F \subseteq H$

or $H \subseteq F$. A subfunctor of a functor, or a submodule of a module with such property is called a *waist*, in joint work with E. L. Green [11, 12].

Maurice was excited about the work on almost split sequences and ir- reducible maps from the beginning, but realized that it could be hard to convince others to take it seriously. At conferences he was more inclined to lecture on newer things, even if there was less chance of the material being accepted. For example in connection with discussing what we should lecture about at the 1974 Ottawa meeting he wrote the following, where the first part refers to the work on almost split sequences.

> (March 3, 1974) *If you think this is a good idea, I would suggest either an exposition of the material in our notes perhaps enhanced by what we succeed in getting before then or else a good exposition of the stuff on stable equivalence, something that is really impossi- ble to accomplish in an hour. Of course the stable equivalence stuff is safer since it is done and I'd be perfectly happy to talk about it. Or I'm perfectly happy to try the other course if you are game even though it's a little more risky... The other could be "Almost splitable sequences".*

Not surprisingly we ended up with the "risky" choice.

At the same time Maurice was pleased about some positive reactions to our work, during the spring of 1974, as the following citations show.

> (March 22, 1974) *Reiner also claimed that the whole thing is begin- ning, just beginning I fear, to make sense to him. At Northwest- ern Matlis seemed to be taking a genuine interest. At Chicago, MacLane and Alperin both seemed to like the work, especially Alperin who really seems to be developing a genuine interest.*

Maurice was interested in finding connections in mathematics, but even for him there were limits for what he thought could be connected, and here he was also often right.

> (March 22, 74) *Szpiro came to Urbana while I was there and it was nice seeing him again. Peskine and he still feel I am doing this work in order to get at the Tor conjecture. I told him this was not the case, but I don't think I convinced him. I guess it would really be funny if I went back to that problem and solved it, especially if some of these ideas were involved. But I don't think things will work out that way.*

In general Maurice was also concerned about how the material was presented, and especially in connection with the work on almost split sequences and irreducible maps, which he found mysterious, and assumed would look even more mysterious to others.

(Aug. 12, 1975) *Hope to be back at work next week, including the manuscripts IV and V. Part of the problem is that I am dissatisfied, not so much with what you have written, but with the fact that almost split sequences still remain such a mystery. In particular I'm annoyed that no better proof of their existence has appeared. But I'm convinced now that that will have to wait.*

(Sept. 10, 75) *In its present form I find it quite startling and unexpected. It should have some interesting applications eventually, but I've not found any as yet. This isn't too surprising since the whole thing is so new. Do you see what to do with it. The subject seems to have no end of unexpected results.*

(Sept. 12, 1975) *I did not include the functorial material in my revised IV since it wasn't necessary and might frighten people off from looking at the rest of the material. You developed, worked with and have generally lived with these techniques for a long time so to you they are routine calculations, but to some one seeing them for the first time I'm afraid the whole thing will be an incomprehensible mystery. The subject is already mysterious enough to everyone without being added by hasty writing.*

As apparent from the citations, Maurice was not completely happy about the original existence proofs. His favourite proof, which is the one given in the book [39], is based on the following property of $D\,\mathrm{Tr}$.

Proposition 2.9 *Let* $0 \to A \xrightarrow{f} B \xrightarrow{g} C \to 0$ *be an exact sequence in* $\mathrm{mod}\,\Lambda$, *where* Λ *is an artin algebra, and let* X *be in* $\mathrm{mod}\,\Lambda$. *Then any map* $s : X \to C$ *factors through* $g : B \to C$ *if and only if any map* $t : A \to D\,\mathrm{Tr}\,X$ *factors through* $f : A \to B$.

This property is based on the following formula which Maurice was very fond of, and which was the basis of our study of indecomposable modules determined by their composition factors, via the concept of short chain [21].

Proposition 2.10 *Let* A *and* B *be in* $\mathrm{mod}\,\Lambda$ *for an artin R-algebra* Λ, *and let* $P_1 \to P_0 \to A \to 0$ *be a minimal projective presentation of* A. *Denoting the length of* $\mathrm{Hom}_\Lambda(X,Y)$ *as an R-module by* $\langle X,Y \rangle$ *we have the formula*

$$\langle A, B \rangle - \langle B, D\,\mathrm{Tr}\,A \rangle = \langle P_0, B \rangle - \langle P_1, B \rangle.$$

The notion of an indecomposable module B being the middle of a short chain, which we defined to mean that there is some indecomposable module A with $\langle A, B \rangle \neq 0$ and $\langle B, D \operatorname{Tr} A \rangle \neq 0$ was suggested to us by this formula. We gave an application to the problem when indecomposable modules are determined by their composition factors [21]. This result has later been improved.

Proposition 2.11 *If A and B are indecomposable modules over an artin algebra Λ which are not the middle of a short chain, then A and B are isomorphic if they have the same composition factors including multiplicity.*

In order to make almost split sequences more interesting to people working in other parts of algebra, Maurice was often working on existence proofs in various contexts, like group representation theory or commutative algebra, using terminology and properties familiar to practitioners in these areas.

Since an almost split sequence $0 \to A \to B \to C \to 0$ is uniquely determined by the module C up to isomorphism, for example the number of summands in a direct sum decomposition of B in indecomposable modules is an invariant of C, usually called $\alpha(C)$. And $\beta(C)$ denotes the number of the summands which are not projective. These invariants give information on the AR-quiver (Auslander-Reiten quiver), which is a quiver associated with an artin algebra based on the information provided by almost split sequences and irreducible maps. We investigated $\alpha(C)$ in [19], and obtained further information on these invariants through our study of uniserial functors [20]. For example we showed that if Λ is of finite representation type and $\beta(C) \leq 2$ for all indecomposable nonprojective modules C, then Λ is biserial, that is if P is indecomposable projective and not uniserial then either $\underline{r}P$ or $\underline{r}P/\operatorname{soc} P$ is a direct sum of two uniserial modules. In particular this gave necessary conditions for an algebra to be stably equivalent to a Nakayama algebra.

Already in the mid seventies Maurice extended our existence theorem for almost split sequences over artin algebras to a much more general higher dimensional setting, containing (maximal) Cohen-Macaulay modules over complete local commutative Cohen-Macaulay rings. The left end of an almost split sequence could be expressed in terms of the right end by an operation τ analogous to $D \operatorname{Tr}$ for artin algebras. Maurice was very excited about this development, and he wrote the following, before having the most general setting and an analogue of $D \operatorname{Tr}$.

(July 16, 1975) *It is really amazing how connected mathematics is. I never thought that the work on artin algebras of finite representation type would lead to results about complete noetherian local rings in general. I find the situation very interesting and exciting. If an analogue existed for the dual of the transpose, it would be most interesting.*

Maurice kept his interest in almost split sequences to his last days. He was looking for more basic properties, and he obtained some curious results on homological properties of almost split sequences which he was working on writing up when he died.

3   Commutative rings

Even though Maurice's general existence theorem for almost split sequences was available in the mid seventies, it was not until around 1983 that he found his important and influential applications to (maximal) Cohen-Macaulay modules over commutative rings and singularity theory. Also in the general setting there is the concept of finite representation type, which in the commutative case expresses that there is only a finite number of indecomposable (maximal) Cohen-Macaulay modules up to isomorphism. In addition there is an associated AR-quiver, and Maurice discovered an interesting connection between the AR-quivers of two-dimensional invariant rings $k[[X, Y]]^G$ where G is a finite subgroup of $SL(2, k)$, for an algebraically closed field k of characteristic zero, and the resolution graph of the corresponding isolated singularity. Maurice's work went via the associated McKay quiver, and through this also the Dynkin and extended Dynkin diagrams appeared again. An important feature is that the operation τ expressing the left end of an almost split sequence in terms of the right end is the identity in this setting. This work opened up a new area of research, and most of my work with Maurice in the following years dealt with commutative ring theory, and also higher dimensional noncommutative ring theory, inspired by this connection.

When the characteristic of the field is zero, the invariant rings $k[[X, Y]]^G$ with G a finite subgroup of $SL(2, k)$ are exactly the two-dimensional commutative Gorenstein rings (which are complete local with residue field k) of finite representation type, that is having only a finite number of indecomposable (maximal) Cohen-Macaulay modules up to isomorphism. The inclusion $G \subset SL(2, k)$ determines a representation of G over k, which again determines a quiver, called the McKay quiver, whose vertices correspond to the irreducible representations of G over k. It was observed by McKay that these quivers are of the form $\tilde{\Delta}$, obtained from an extended Dynkin diagram Δ by replacing each edge $\cdot - \cdot$ by a pair of arrows $\cdot \rightleftarrows \cdot$. McKay further observed that deleting the vertex corresponding to the trivial representation k we have $\tilde{\Delta}_0$ where Δ_0 is a Dynkin diagram, and is the resolution graph of the corresponding isolated singularity. From the isomorphism between the AR-quiver for $k[[X, Y]]^G$ and the McKay quiver it follows that the AR-quiver is also $\tilde{\Delta}$ where Δ is extended Dynkin. When the vertex corresponding to the ring is deleted, the AR-quiver is $\tilde{\Delta}_0$ where Δ_0 is the Dynkin diagram of the isolated singularity. In arbitrary characteristic the Gorenstein rings of finite representation type are the rational double points, and the corresponding resolution graphs are still Dynkin. Artin and Verdier had already established a direct connection between the indecomposable nonprojective (maximal) Cohen-Macaulay modules and the edges of the singularity graph. In [24] we proved the following.

Theorem 3.1 *Let Λ be a rational double point over an algebraically closed field k. Then the AR-quiver of Λ is of the form $\tilde{\Delta}$ where Δ is an extended Dynkin diagram and of the form $\tilde{\Delta}_0$ with Δ_0 Dynkin if the vertex corresponding to Λ is deleted.*

Using amongst other things this result, Esnault and Knörrer found a direct connection between the resolution graph and the AR-quiver with the vertex corresponding to the ring deleted.

For an algebraically closed field k, the complete local commutative two-dimensional Cohen-Macaulay rings Λ (with residue field k) of finite representation type are exactly the invariant rings $k[[X, Y]]^G$ where G is a finite subgroup of $GL(2, k)$. One problem was to find commutative rings of finite representation type in dimension greater than two when the ring was not a hypersurface. The hypersurface case in characteristic different from two had been dealt with, and the characteristic two case was completed later. We found two such examples, each of them being the only one of finite representation type in a class to which it naturally belongs [25].

Theorem 3.2 *Let k be an algebraically closed field and G a finite subgroup of $GL(n, k)$ for some $n \geq 3$. Then the invariant ring $k[[X_1, \ldots, X_n]]^G$ is of finite representation type if and only if $n = 3$ and $G = \left\langle \begin{pmatrix} -1 & 0 & 0 \\ 0 & -1 & 0 \\ 0 & 0 & -1 \end{pmatrix} \right\rangle$.*

Recall that a scroll of type $(n_1, \ldots, n_r)$ is a factor of the power series ring $k[[X_0^{(1)}, \ldots, X_{n_1}^{(1)}, \ldots X_0^{(n)}, \ldots, X_{n_r}^{(n)}]]$ modulo the ideal generated by the determinants of the 2×2 minors of the matrix $\begin{pmatrix} X_0^{(1)} & X_{n_1-1}^{(1)} \\ X_1^{(1)} & X_{n_1}^{(1)} \end{pmatrix} \cdots \begin{pmatrix} X_0^{(r)} \cdots X_{n_r-1}^{(r)} \\ X_1^{(r)} \cdots X_{n_r}^{(r)} \end{pmatrix}$. It is then known that $\dim \Lambda = t + 1$.

Theorem 3.3 *Let Λ be a scroll of type $(n_1, \ldots, n_t)$ with $t \geq 2$. Then Λ is of finite representation type if and only if it is of type $(2, 1)$.*

For proving finite representation type we used a criterion based on the existence of almost split sequences, but which is not a formal consequence of the existence. Such a criterion was proved by Maurice for artin algebras and had already been an important method for proving finite representation type in that setting. It essentially amounts to making a guess for a finite set of indecomposable objects which are believed to be all and then show that one can build almost split sequences for these modules just using the modules from the finite set [25].

For the invariant rings we also gave an alternative proof for finite representation type without using almost split sequences. In proving infinite representation type we were able to reduce the problem to showing that certain artin algebras with radical square zero are of infinite representation type, for which there is a nice criterion as discussed in section 1. This represented a different type of connection with artin algebras.

These new rings of finite representation type we found around 1984. After hearing about the example of the invariant ring, Eisenbud suggested during my stay at Brandeis to look at the class of scrolls. We tried for a while to prove that all or more of them were of finite representation type, but finally started suspecting that they were no more.

Maurice had proved that almost split sequences exist in the category of (maximal) Cohen-Macaulay modules for Λ if and only if Λ is an isolated singularity. Inspired by a question of Schreyer in 1985 we proved the following local version of this result [28].

Theorem 3.4 *Let R be a complete local commutative Gorenstein ring and Λ an R-algebra which is a finitely generated (maximal) Cohen-Macaulay R-module. Let C be indecomposable nonprojective in the category $\mathrm{CM}(\Lambda)$ of Λ-modules which are (maximal) Cohen-Macaulay modules over R.*

Then there is an almost split sequence $0 \to A \to B \to C \to 0$ in $CM(\Lambda)$ if and only if C_p is a free R_p-module for each maximal prime ideal p in R.

We also investigated existence of almost split sequences for graded Cohen-Macaulay modules over graded Cohen-Macaulay rings, partially motivated by a possible application to existence of almost split sequences for vector bundles and coherent sheaves. We first dealt with the twodimensional case [26]. We obtained a local version also in the $\mathbb{Z}$-graded case [27].

Theorem 3.5 *Let $R = k[X_1, \ldots, X_d]$ be a polynomial ring over a field k together with a $\mathbb{Z}$-grading such that the degrees of the X_i are at least one and the constants have degree zero. Let S be a $\mathbb{Z}$-graded R-algebra which is a finitely generated projective R-module, and denote by $\mathrm{CM}(\mathrm{gr}S)_0$ the category whose objects are the finitely generated graded S-modules which are projective R-modules and where the morphisms are of degree zero. Let C be an indecomposable projective in $\mathrm{CM}(\mathrm{gr}S)_0$.*

Then there is an almost split sequence $0 \to A \to B \to C \to 0$ in $\mathrm{CM}(\mathrm{gr}S)_0$ if and only if C_p is is Λ_p-projective for each nonmaximal prime ideal p in R.

In particular we have the following consequence.

Corollary 3.6 *Almost split sequences exist for vector bundles and for coherent sheaves on nonsingular projective curves.*

The method of proof was based on a graded version of the formula $D \operatorname{Ext}^1(C, \tau B) \xrightarrow{\sim} \underline{\operatorname{Hom}}(B, C)$. In fact, this is an analogue of the Serre duality formula, which alternatively can be used directly to prove existence of almost split sequences for nonsingular projective curves, as done in the independent approach by Schofield.

A graded R-algebra S as above is said to be of finite graded representation type if there is only a finite number of indecomposable graded (maximal) Cohen-Macaulay S-modules up to shift. We imitiated the theory further, in establishing similar criteria for proving finite graded representation type. These criteria were applied it to the graded scroll of type $(2, 1)$ in [25] and to the graded invariant ring $S = k[X_1, X_2, X_3]^G$ by Solberg. These results indicated that there was a connection between graded finite representation type for a graded ring S, and finite representation type for the completion $\Lambda = \hat{S}$ with respect to $(X_1, \dots, X_d)$, and also between the almost split sequences for S and $\Lambda = \hat{S}$. In an effort to explain these features, we were able to prove the following [27].

Theorem 3.7 *Let k be a field, $R = k[X_1, \dots, X_n]$ and S a $\mathbb{Z}$-graded R-algebra, which is finitely generated projective as a R-module.*

(a) *If $0 \to A \to B \to C \to 0$ is an almost split sequence in the category $\operatorname{CM}(\operatorname{gr} S)_0$, then the induced sequence $0 \to \hat{A} \to \hat{B} \to \hat{C} \to 0$ obtained by completing with respect to $(X_1, \dots, X_n)$ is an almost split sequence in $\operatorname{CM}(\hat{S})$.*

(b) *S is of finite graded representation type if and only if $\hat{S}$ is of finite representation type.*

There are also existence theorems for S being graded by more general groups, and then the almost split sequence may go to a direct sum of almost split sequences via completion [33].

There is an interesting connection between rational double points R for an algebraically closed field k and finite dimensional k-algebras. For if M denotes a direct sum of one copy of each indecomposable object in the category of (maximal) Cohen-Macaulay modules $\operatorname{CM}(R)$, then the endomorphism algebra $\Gamma = \operatorname{End}(M)^{\operatorname{op}}$ is finite dimensional over k since R is an isolated singularity. And using the shape of the AR-quiver one can prove that Γ is the preprojective algebra Π of the corresponding Dynkin diagram Δ. Recall that Π is the path algebra of $\tilde{\Delta}$ over k modulo the ideal generated by the elements r_i for each vertex in Δ, defined by $r_i = \sum_\alpha \alpha^* \alpha$, where α runs through the arrows starting at i and α^* is the arrow obtained from α by changing direction.

Note that as for artin algebras the stable category $\underline{CM}(R)$ is equivalent to the category of projective modules in $\text{mod}\,\Gamma$. In [40] we applied this connection to study the module theory of the preprojective algebra of a Dynkin diagram. We were able to use the fact that for R the correspondence τ between the end terms of an almost split sequence is the identity and that the functor $\Omega^2 : \underline{\text{mod}}R \to \underline{\text{mod}}R$ is isomorphic to the identity to show the following, proved by Ringel and Schofield using different methods [40].

Proposition 3.8 *Let Γ be the preprojective algebra of a Dynkin diagram. Then for each indecomposable nonprojective module in* $\text{mod}\,\Gamma$ *we have* $(D\,\text{Tr})^6 C \simeq C$.

Most of the preprojective algebras are of wild representation type. In fact, applying the same method to commutative rings of dimension one we obtained the following [40].

Proposition 3.9 *There is a one-dimensional Gorenstein ring R of finite representation type such that if M is the direct sum of one copy of each indecomposable module in $CM(R)$, then $\Gamma = \underline{\text{End}}(M)^{\text{op}}$ is a finite dimensional wild algebra where $(D\,\text{Tr})^3 C \simeq C$ for each indecomposable nonprojective module C in* $\text{mod}\,\Gamma$.

Our work in [40] was formulated in a more general setting, dealing with functor categories in a similar way as we discussed for artin algebras in section 1. An interesting feature is that when R is an isolated singularity where almost split sequences exist, then $\underline{CM}(R)$ is a dualizing R-variety, by a slight extension of the definition, allowing R to be a noetherian ring [40]. In fact it is the analogue of the formula in Theorem 2.4 which is used to prove existence of almost split sequences, that essentially expresses the fact that $\underline{CM}(R)$ is a dualizing R-variety. Then it makes sense to talk about almost split sequences in the category of finitely presented contravariant functors from $\underline{CM}(R)$ to abelian groups.

We also worked on Grothendieck groups (with relations given by all exact sequences), in particular for the invariant rings $\Lambda = k[[X_1, \dots, X_n]]^G$ when G is a finite subgroup of $\text{GL}(n, k)$. Our method was to use the skew group ring $k[[X_1, \dots, X_n]]G$ and long exact sequences of K-groups. I state one of our results, but not in complete generality [22, 31].

Theorem 3.10 *Let $\sum$ be a complete regular local domain and G a finite group acting on $\sum$ as ring automorphisms such that the order of G is invertible in $\sum$. Then the Grothendieck group $K_0(\text{mod}\,\sum^G)$ is isomorphic to $\mathbb{Z} \oplus H$ where H is a finite group.*

There is an interesting connection between Grothendieck groups and almost split sequences, and for finite representation type it is actually possible to compute the Grothendieck group from the AR-quiver [22]. This is based upon a generalization of a result by Butler for artin algebras.

4 Tilting theory

Bernstein, Gelfand and Ponomarev gave in the early seventies a different approach to the theorem of Gabriel classifying the quivers for which there is only a finite number of indecomposable representations up to isomorphism over a field k. They used certain reflection functors, defined from the category of representations of a quiver without oriented cycles to the category of representations of a related quiver. This was generalized to representations of species by Dlab and Ringel.

For example if we have the quiver $\begin{smallmatrix}1\\\downarrow\alpha\\2\\{}_{\gamma}\end{smallmatrix}$ and fix the vertex 4, let Γ' be the quiver $\begin{smallmatrix}1\\\downarrow\alpha\\2\end{smallmatrix}$. To a representation $\begin{smallmatrix}V_1\\\downarrow f_\alpha\\V_2\\{}_{f_\gamma}\end{smallmatrix}$ of Γ we associate the representation $\begin{smallmatrix}V_1\\\downarrow f_\alpha\\V_2\\{}_{f_{\gamma'}}\end{smallmatrix}$ of Γ', where $f_{\gamma'} : V_4' \to V_2$ is determined by the exact sequence $0 \to V_4' \xrightarrow{f_{\gamma'}} V_2 \xrightarrow{f_\gamma} V_1$.

Since the reflection functors were important for hereditary algebras we hoped that similar functors would be useful more generally. We started by trying to understand the reflection functor from a module theoretic point of view, together with Platzeck, during our stay at Brandeis 1976-77. We proved the following first module theoretic version of what is now called tilting theory [18].

Theorem 4.1 *Let Λ be a basic artin algebra with a simple projective noninjective module S. Write $\Lambda = S \amalg P$ and let $T = P \amalg \operatorname{Tr} D(S)$.*

(a) *The functor $\operatorname{Hom}_\Lambda(T, \ ) : \operatorname{mod}\Lambda \to \operatorname{mod}\operatorname{End}_\Lambda(T)^{\mathrm{op}}$ is a fully faithful functor when restricted to the full subcategory $\operatorname{mod}_S \Lambda$ of $\operatorname{mod}\Lambda$ whose objects do not have S as a summand.*

(b) *If in addition $\operatorname{Hom}_\Lambda(\operatorname{Tr} DS, \Lambda) = 0$, then $S' = \operatorname{Ext}^1_\Lambda(T, S)$ is a simple injective $\operatorname{End}_\Lambda(T)^{\mathrm{op}}$-module and $\operatorname{Hom}_\Lambda(T, \ )$ induces an equivalence $\operatorname{Hom}_\Lambda(T, \ ) : \operatorname{mod}_S \Lambda \to \operatorname{mod}_{S'} \operatorname{End}_\Lambda(T)^{\mathrm{op}}$.*

This work was also inspired by almost split sequences. Actually the almost split sequence $0 \to S \to E \to \operatorname{Tr} DS \to 0$ played an essential role. But it has later turned out that one does not need almost split sequences here. Shortly after finishing the work we had the chance to present it at the 1977 Oberwolfach meeting. The work created some interest and was further generalized and developed into what is now called tilting theory, which plays a central role in the representation theory of artin algebras.

A tilting Λ-module T is a Λ-module satisfying the following:

(i) $\operatorname{pd}_\Lambda T \leq 1$ (where $\operatorname{pd}_\Lambda T$ denotes the projective dimension of T).

(ii) $\operatorname{Ext}^1_\Lambda(T, T) = 0$

(iii) There is an exact sequence $0 \to \Lambda \to T_0 \to T_1 \to 0$ with T_0 and T_1 in add T, that is T_0 and T_1 are direct summands of the direct sum of a finite number of copies of T.

The module $T = P \amalg \operatorname{Tr} DS$ in Theorem 4.1 is an example of a tilting module, and is in the literature sometimes called an APR-tilting module.

Our second main contribution to tilting theory came in the late eighties, after the notion of tilting, and the dual notion of cotilting, module had been generalized further.

A Λ-module T is a (generalized) tilting module if

(i) $\operatorname{pd}_\Lambda T < \infty$

(ii) $\operatorname{Ext}_\Lambda{}^i(T, T) = 0$ for all $i > 0$.

(iii) there is an exact sequence $0 \to \Lambda \dashrightarrow T_0 \to \cdots \to T_n \to 0$ with each T_i in add T.

Dually T' is a (generalized) cotilting module if $D(T')$ is a tilting Λ^{op}-module.

In [34, 37] we proved a connection between (generalized) cotilting modules and certain contravariantly finite subcategories of mod Λ. The theory of contravariantly (and covariantly) finite subcategories goes back to Maurice's work with Smalø. We recall that a subcategory $\mathcal{C}$ of mod Λ, closed under isomorphisms and summands, is *contravariantly finite* if for each X in mod Λ there is some map $f : C \to X$ with C in $\mathcal{C}$, such that for any map $g : C' \to X$ with C' in $\mathcal{C}$ there is some $h : C' \to C$ such that $fh = g$. The dual notion is that $\mathcal{C}$ is *covariantly finite*. Further we recall that $\mathcal{C}$ is *resolving* if it contains all projective modules and is closed under extensions and kernels of epimorphisms. Finally, for a full subcategory $\mathcal{C}$ of mod Λ denote by $\widehat{\mathcal{C}}$ the full subcategory of mod Λ whose objects are the X for which there is an exact sequence $0 \to C_n \to \cdots \to C_1 \to C_0 \to X \to 0$.

Theorem 4.2 *Let Λ be an artin algebra and let $\mathcal{X}$ be a full subcategory of* mod Λ.

(a) *If $\mathcal{X}$ is a contravariantly finite resolving subcategory of* mod Λ *with* $\widehat{\mathcal{X}} = \text{mod } \Lambda$, *then T is a cotilting module if* add T *is the Ext-projectives in $\mathcal{X}$.*

(b) *If T is a cotilting module, then $\mathcal{X}_T = \{C; \operatorname{Ext}_\Lambda{}^i(C,T) = 0; i > 0\}$ is a contravariantly finite resolving subcategory of* $\operatorname{mod}\Lambda$ *with* $\widehat{\mathcal{X}}_T = \operatorname{mod}\Lambda$.

The above correspondence is especially nice for algebras of finite global dimension, since $\widehat{\mathcal{X}} = \operatorname{mod}\Lambda$ is then automatically satisfied for a resolving subcategory $\mathcal{X}$ of $\operatorname{mod}\Lambda$. Then we get a one-one correspondence between contravariantly finite subcategories and (basic) cotilting modules. A beautiful illustration was made by Ringel for quasihereditary algebras, and there were further applications to algebraic groups and quantum tilting.

When T is a cotilting module with $\operatorname{id}_\Lambda T \leq 1$, where $\operatorname{id}_\Lambda T$ denotes the injective dimension of T, then the above category $\mathcal{X}_T$ is $\operatorname{Sub} T$, that is the objects are submodules of finite direct sums of copies of T. Already in Maurice's work with Smalø a connection between the subcategories of the form $\operatorname{Sub} T$ and classical cotilting modules was observed. Our work was to a large extent also inspired by Maurice's work on Cohen-Macaulay approximations with Buchweitz. In this case the dualizing module for a Cohen-Macaulay ring plays the role of a cotilting module, and also has analogous properties [34]. This connection motivated a study of a class of artin algebras which we called Cohen-Macaulay algebras [35].

5 Contravariantly finite subcategories and syzygies

We saw in the previous section there that is an interesting connection between contravariantly finite subcategories of $\operatorname{mod}\Lambda$ and cotilting modules for an artin algebra Λ. The contravariantly finite subcategories are also important in connection with almost split sequences, since Maurice had shown in joint work with Smalø that subcategories (closed under isomorphisms and summands) which are both contravariantly and covariantly finite and closed under extensions have almost split sequences. These facts motivate investigating general properties of contravariantly (or covariantly) finite subcategories and procedures for constructing such subcategories. It is then also important to find examples of contravariantly and/or covariantly finite subcategories which are in addition resolving (or coresolving), or at least extension closed. We worked on such questions since the late eighties, and we were here also influenced by Maurice's earlier work on commutative and noncommutative ring theory, in particular by his work with Bridger.

The following gives a way of constructing covariantly finite subcategories from contravariantly finite ones and conversely [34].

Theorem 5.1 *Let $\mathcal{X}$ be a contravariantly finite extension closed subcategory of* mod Λ *for an artin algebra* Λ.

(a) *Then* $\mathcal{Y} = \{C; \mathrm{Ext}^1(\mathcal{X}, C) = 0\}$ *is covariantly finite extension closed and contains the injectives, and* $(\mathcal{Z} = \{C; \mathrm{Ext}^1(C, \mathcal{Y}) = 0\}$ *is contravariantly finite, extension closed and contains the projectives.*

(b) *If* $\mathcal{X}$ *contains the projectives, then* $\mathcal{Z} = \mathcal{X}$.

The contravariantly finite subcategories of mod Λ which are in addition resolving are "finitely generated" in the following sense.

Theorem 5.2 *Let $\mathcal{X}$ be a contravariantly finite resolving subcategory of* mod Λ *for an artin algebra* Λ. *Denote by* $S_1, \dots, S_n$ *the simple* Λ-*modules up to isomorphism and let* $h_i : X_{S_i} \to S_i$ *be a minimal right* $\mathcal{X}$-*approximation of* S_i. *Then each* X *in* $\mathcal{X}$ *is a summand of a module which has a filtration with factors amongst the* X_{S_i}.

There is a curious application of this fact to a partial result for the finitistic dimension conjecture [34].

Corollary 5.3 *If* $\mathcal{C} = \{C; \mathrm{pd}_\Lambda C < \infty\}$ *is contravariantly finite, then* fin dim $\Lambda < \infty$.

Recall that the finitistic dimension conjecture says that $\sup\{\mathrm{pd}\, C; \mathrm{pd}\, C < \infty\}$ is finite. A consequence of this conjecture is the Nakayama conjecture, which says that if in a minimal injective resolution $0 \to \Lambda \to I_0 \to I_1 \to \cdots \to I_j \to \cdots$ each I_j is projective, then Λ is selfinjective. In [10] we studied a more general question, which we called the generalized Nakayama conjecture: If Λ is an artin algebra with a resolution as above and I is an indecomposable injective Λ-module, then I is a summand of I_j for some j. These conjectures are still unsolved in general.

Partially motivated by looking for interesting classes of examples for contravariantly finite (resolving) subcategories, we started to investigate syzygy categories. For some of this work we only assumed Λ to be a twosided noetherian ring. For each $i > 0$ denote by $\Omega^i(\mathrm{mod}\,\Lambda)$ the full subcategory of mod Λ whose objects are the i-th syzygy modules, that is those C for which there is an exact sequence

$$0 \to C \to P_{i-1} \to \cdots \to P_1 \to P_0 \to X \to 0$$

with the P_i projective. We showed that $\Omega^i(\mathrm{mod}\,\Lambda)$ is always covariantly finite in mod Λ [37]. For artin algebras it is also contravariantly finite as shown for algebras over a field in [36], and extended to artin algebras by Smalø (see [38]). By definition $\Omega^i(\mathrm{mod}\,\Lambda)$ contains the projectives, and we clearly have $\Omega^i(\widehat{\text{mod}\Lambda}) = \text{mod}\,\Lambda$. However $\Omega^i(\text{mod}\,\Lambda)$ is not in general extension closed, actually not even closed under summands. But if it is extension closed, then it is resolving [37]. Hence we were led to investigate when $\Omega^i(\text{mod}\,\Lambda)$ is extension closed, and we did this for an arbitrary noetherian ring Λ. Note that when $\Omega^i(\text{mod}\,\Lambda)$ is extension closed for an artin algebra Λ, then the category $\Omega^i(\text{mod}\,\Lambda)^{\perp}$, which by Theorem 5.1 is covariantly finite, is $\{C; \text{id}C \le i\}$.

The first general sufficient condition we obtained for $\Omega^i(\text{mod}\,\Lambda)$ to be extension closed was that Λ belongs to a class of rings which Maurice introduced in the late sixties, while looking for an appropriate noncommutative analogue of commutative Gorenstein rings. For an integer $d \ge 1$ he defined Λ to be d-Gorenstein if in a minimal injective resolution $0 \rightarrow \Lambda \rightarrow I_0 \rightarrow I_1 \rightarrow \cdots \rightarrow I_i \rightarrow \cdots$ we have flat dim $I_i \le i$ for all $i < d$. The noncommutative analogue for Gorenstein was that Λ is d-Gorenstein for all d, and such a Λ is now called a ring satisfying the Auslander conditions. It turned out that these sufficient conditions could be generalized, and after various stages of improvements we were able to prove the following. Note that the extra assumption of noetherian R-algebra is only used for the implication (a) $\Rightarrow$ (c) when $d > 1$.

Theorem 5.4 *Let Λ be a noetherian algebra over a commutative noetherian ring R and $d \ge 1$ an integer. Then the following are equivalent.*

(a) $\Omega^i(\text{mod}\,\Lambda)$ *is extension closed for $i \le d$.*

(b) *If $0 \rightarrow \Lambda \rightarrow I_0 \rightarrow I_1 \rightarrow \cdots \rightarrow I_i \rightarrow \cdots$ is a minimal injective resolution of Λ as a right Λ-module, then flat dim $I_i \le i+1$ for $i < d$.*

(c) grade $\text{Ext}_\Lambda^i(C, \Lambda) \ge i$ *for all C in $\text{mod}\,\Lambda^{\text{op}}$ and $i \le d$.*

(d) grade $X \ge i$ *for $X \subset \text{Ext}_\Lambda^{i+1}(B, \Lambda)$ when B is in $\text{mod}\,\Lambda$ and $i \le d$.*

The final proof is given in [41], which was our last joint paper, being typed while I accompanied Maurice on his nostalgic tour to Europe a few weeks before he died.

The related category $J_d(\text{mod}\,\Lambda) = \text{Tr}\,\Omega^d(\text{mod}\,\Lambda^{\text{op}})$, which played an important role in Maurice's work with Bridger from the sixties is contravariantly finite for a noetherian ring Λ, but is not in general covariantly finite. There is an interesting connection with (maximal) Cohen-Macaulay modules if Λ is a local commutative noetherian Gorenstein ring of dimension d. Then $\Omega^d(\text{mod}\,\Lambda) = J_d(\text{mod}\,\Lambda) = CM(\Lambda)$, accounting for the fact that $CM(\Lambda)$ is both contravariantly and covariantly finite in $\text{mod}\,\Lambda$. We also discovered some curious connections between $J_d(\text{mod}\,\Lambda)$ and categories of modules of projective dimension at most i for some i. We worked on this in connection with preparing for the 1994 Utrecht meeting on Cohen-Macaulay modules, Cohen-Macaulay approximations and singularity theory. This work is still to be written up.

We had also started investigating noetherian rings of global dimension two from the point of view of properties of the transpose and the modules of grade two. This was the topic of Maurice's last public lecture, in Mexico at ICRA VII, and this joint work is written up in [42].

Almost all my joint work with Maurice had some connection to one of the five headlines I have chosen. One paper which did not fit was on work with Smalø on Galois actions, where we tried to understand some of the covering theory from the point of view of rings.

I feel I was fortunate to have such an extensive and close collaboration with Maurice. Working with him was exciting and stimulating, and his true love for mathematics and its beauty had a strong influence on me. The loss of a close collaborator and even more of a special friend, is difficult.

References

[1]   M. Auslander and I. Reiten, *Representation theory of artin algebras*, Brandeis Univ. 1972.

[2]   M. Auslander and I. Reiten, *Stable equivalence of artin algebras*, Proc. Conf. on Orders, Group rings and related topics (Ohio 1972), Lecture Notes in Math. 353, Springer-Verlag (1973) 8–71.

[3]   M. Auslander and I. Reiten, *Stable equivalence of dualizing R-varieties*, Adv. in Math., Vol. 12, No. 3 (1974) 306–366.

[4]   M. Auslander and I. Reiten, *Stable equivalence of dualizing R-varieties II: Hereditary dualizing R-varieties*, Adv. in Math., Vol. 17, No. 2 (1975) 93–121.

[5]   M. Auslander and I. Reiten, *Stable equivalence of dualizing R-varieties III: Dualizing R-varieties stably equivalent to hereditary dualizing R-varieties*, Adv. in Math., Vol. 17, No. 2 (1973) 122–142.

[6]   M. Auslander and I. Reiten, *Stable equivalence of dualizing R-varieties IV: Higher global dimension*, Adv. in Math., Vol. 17, No. 2 (1975) 143–166.

[7]   M. Auslander and I. Reiten, *Stable equivalence of dualizing R-varieties V: Artin algebras stably equivalent to hereditary algebras*, Adv. in Math. Vol. 17, No. 2 (1975) 167–195.

[8]   M. Auslander and I. Reiten, *Representation theory of artin algebras III: Almost split sequences*, Comm. in Algebra, Vol. 3, No. 3 (1975) 239–294.

[9]   M. Auslander and I. Reiten, *Almost split sequences II*, Proc. int. conf. on representation of algebras, Ottawa 1974, Lectore Notes in Math. 488, Springer-Verlag (1975) 9–19.

[10] M. Auslander and I. Reiten, *On a generalized version of the Nakayama conjecture*, Proc. Amer. Math. Soc., Vol. 52 (1975) 69–74.

[11] M. Auslander, E. L. Green and I. Reiten, *Modules having waists*, Proc. int. conf. on representations of algebras, Ottawa 1974, Lecture Notes in Math. 488, Springer-Verlag (1975) 20–28.

[12] M. Auslander, I. Reiten and E. L. Green, *Modules with waists*, Illinois J. Math. (1975) 467–477.

[13] M. Auslander and I. Reiten, *On the representation type of triangular matrix rings*, J. of London Math. Soc., Vol. 2, No. 12 (1976) 371–382.

[14] M. Auslander and I. Reiten, *Representation theory of artin algebras IV: Invariants given by almost split sequences*, Comm. in Algebra, 5 (1977) 443–518.

[15] M. Auslander and I. Reiten, *Representation theory of artin algebras V: Methods for computing almost split sequences and irreducible morphisms*, Comm. in Algebra, 5 (1977) 519–554.

[16] M. Auslander and I. Reiten, *Representation theory of artin algebras VI: A functorial approach to almost split sequences*, Comm. in Algebra, 11 (1977) 279–299.

[17] M. Auslander, M. I. Platzeck and I. Reiten, *Periodic modules over weakly symmetric algebra*, Journal of Pure and Applied Algebra 11 (1977) 279–291.

[18] M. Auslander, M. I. Platzeck and I. Reiten, *Coxeter functors without diagrams*, Trans. Amer. Math. Soc. 250 (1976) 1–46.

[19] M. Auslander, R. Bautista, M. I. Platzeck, I. Reiten and S. O. Smalø, *Almost split sequences whose middle term has at most two indecomposable summands*, Canadian J. Math., Vol. XXXI, No. 5 (1979) 942–960.

[20] M. Auslander and I. Reiten, *Uniserial functors*, Proc. ICRA II (1979), Lecture Notes in Math. 832 (1980) 1–47.

[21] M. Auslander and I. Reiten, *Modules determined by their composition factors*, Ill. J. Math., Vol. 29, No. 2 (1985) 280–301.

[22]   M. Auslander and I. Reiten, *Grothendieck groups of algebras and orders*, Journal of Pure and Applied Algebra 39 (1986) 1–51.

[23]   M. Auslander and I. Reiten, *McKay quivers and extended Dynkin diagrams*, Trans. Amer. Math. Soc., Vol. 293, No. 1 (1986) 293–301.

[24]   M. Auslander and I. Reiten, *Almost split sequences for rational double points*, Trans. Amer. Math. Soc., Vol. 302, No. 1 (1987) 87–97.

[25]   M. Auslander and I. Reiten, *The Cohen-Macaulay type of Cohen-Macaulay rings*, Adv. in Math., No. 1 (1989) 1–23.

[26]   M. Auslander and I. Reiten, *Almost split sequences in dimension two*, Adv. in Math., Vol. 66, No. 1 (1987) 88–118.

[27]   M. Auslander and I. Reiten, *Almost split sequences for $\mathbb{Z}$-graded rings*, Proc. of 1985 Lambrecht Conf., Springer Lecture Notes 1273 (1987) 232–243.

[28]   M. Auslander and I. Reiten, *Almost split sequences for Cohen-Macaulay modules*, Math. Annales 277 (1987) 345–349.

[29]   M. Auslander and I. Reiten, *Almost split sequences for abelian group graded rings*, J. Algebra, Vol. 114, No. 1 (1988) 29–39.

[30]   M. Auslander and I. Reiten, *Cohen-Macaulay modules for graded Cohen-Macaulay rings and their completions*, Proc. of 1987 Berkeley Conf. on Commutative Algebras, Math. Sc. Research Inst. Publ.!5 Springer-Verlag (1989) 21–31.

[31]   M. Auslander and I. Reiten, *Grothendieck groups of algebras with nilpotent annihilators*, Proc. Amer. Math. Soc. (1988) 1022–1024.

[32]   M. Auslander, I. Reiten and S. O. Smalø, *Galois actions on rings and finite Galois coverings*, Math. Scand. 65 (1989) 5–32.

[33]   M. Auslander and I. Reiten, *Graded modules and their completions*, Banach Center Publ., Vol. 26, Part 1, PWN-Polish Scientific Publishers, Warsaw 1990, 181–192.

[34]   M. Auslander and I. Reiten, *Applications of contravariantly finite subcategories*, Adv. in Math., Vol. 86, No. 1 (1991) 111–152.

[35]   M. Auslander and I. Reiten, *Cohen-Macaulay and Gorenstein algebras*, Conf. on representations of groups and algebras, Progress in Mathematics, Vol. 95, Birkhäuser, 1991 221–245.

[36] M. Auslander and I. Reiten, *On a theorem of E. Green on the dual of the transpose*, Proc. Tsukuba Conf. on Repr. of Algebras, CMS Conference Proceedings, Vol. 11 (1991) 55–65.

[37] M. Auslander and I. Reiten, *Homologically finite subcategories*, Proc. Tsukuba Workshop on Repr. of Algebras, London Math. Soc. Lecture Notes 168.

[38] M. Auslander and I. Reiten, *k-Gorenstein algebras and syzygy modules*, J. of Pure and Applied Algebra 92 (1994) 1–27.

[39] M. Auslander, I. Reiten and S. O. Smalø, Representation theory of artin algebras, Cambridge Univ. Press 1995, 423 pages.

[40] M. Auslander and I. Reiten, *D Tr-periodic modules and functors*, Proc. ICRA VIII, Mexico (to appear).

[41] M. Auslander and I. Reiten, *Syzygy modules for noetherian rings*, J. Alg. (to appear).

[42] I. Reiten, *Homological theory of noetherian rings*, Proc. ICRA VIII, Mexico (to appear) (based on joint work with M. Auslander).

Idun Reiten,
Department of Mathematics and Statistics,
Norwegian University of Science and Technology,
Trondheim, Norway

The Development of the
Representation Theory of Finite Dimensional Algebras
1968 – 1975

Claus Michael Ringel

ABSTRACT. The representation theory of finite dimensional algebras has seen a dramatic development in the last 28 years. The foundation of the modern representation theory was laid in the years 1968 – 1975. The aim of this historical survey is to describe the main directions of investigations in these eight years. We will single out eight topics which have been discussed in the years 1968 – 1975 and show their relationship to the present interests. In 1968, there was the solution of the first Brauer-Thrall conjecture. The introduction of the Auslander algebras may be considered as the starting point for a systematic study of module categories. The use of quivers, posets and quadratic forms are now important tools in representation theory. Functorial methods such as Coxeter functors and functorial filtrations of the forgetful functor were introduced during that period in order to deal with specific classification problems. All these methods have turned out to be very fruitful. As we will see, the main emphasis of most investigations was directed towards an understanding of the different representation types: finite, tame and wild, and they were confined to specific classes of algebras. With the proof of the existence of almost split sequences in 1975 Auslander and Reiten presented a result which deals with arbitrary finite dimensional algebras; the notion of an irreducible map and the concept of the corresponding Auslander-Reiten quiver are now basic ingredients of representation theory.

The Setting

Let k be a field, and A a finite dimensional k-algebra (associative, with 1). We consider representations of A, these are algebra homomorphisms from A into the endomorphism algebra of a vector space over k, or, equivalently, (left) A-modules. The A-modules which we will deal with will usually be finite dimensional (there are two exceptions, in our discussion of the Auslander algebras and in the section dealing with the possible representation types, they will be noted explicitly).

1991 *Mathematics Subject Classification.* Primary 16G10, 16G20.

The classical structure theory of finite dimensional k-algebras shows that such an algebra A has a largest nilpotent ideal $J(A)$ (the radical of A), and that the factor algebra $A/J(A)$ is a product of finitely many matrix rings over division algebras. The Morita theory asserts that the category of A-modules is equivalent to the category of A_0-modules, where A_0 is a basic algebra, this means that $A_0/J(A_0)$ is a product of division algebras. Since the main interest of representation theory lies in properties of the category A-mod of A-modules, we can (and will) assume that A is a basic algebra.

We will consider mainly the case when k is an algebraically closed field. Under this assumption, it is now customary to write A as given by a quiver with relations, thus as a factor algebra of the path algebra kQ of a finite quiver Q modulo what is called an admissible ideal (an ideal which is contained in $(kQ^+)^2$, where kQ^+ is the ideal of kQ generated by the arrows). Note that this kind of presentation of A was exhibited explicitly [G2] only during the period which we want to discuss, but the essential ingredients had been known before: it was stressed in the Nagoya papers around 1955 that a finite dimensional algebra over a perfect field can be written as a nice factor algebra of a hereditary algebra (such as the path algebra kQ) and also the concept of a quiver was used for example by Grothendieck a long time ago, under the label of a diagram scheme; it is the merit of Gabriel to have coined the well-suited name 'quiver' and to have stressed its importance as a main working tool for dealing with finite dimensional algebras: 'quiver theory' now often serves as a denomination for the whole modern representation theory of finite dimensional algebras.

The interpretation of the algebra A as the factor algebra of the path algebra of a quiver allows to write the A-modules as representations of the quiver: as given by a finite number of vector spaces V_i and linear maps $V_i \rightarrow V_j$. In this way, the representation theory of algebras turns out to be just a sort of higher linear algebra. Conversely, many classification problems in algebra and geometry can be formulated in terms of vector spaces and linear maps, thus they may be interpreted in terms of representation theory. For example, the classical problem of classifying matrix pencils considered by Weierstraß and Kronecker is just the problem of classifying the representations of the quiver

(now called the Kronecker quiver). The path algebra of this quiver is four dimensional, it can be written as $\begin{bmatrix} k & k^2 \\ 0 & k \end{bmatrix}$.

The aim of the representation theory is to construct and to describe all representations of A, but usually this may be impossible, thus one is hoping to understand at least parts of the module category: on the one hand, one is interested in a description of suitable classes of representations, on the other hand, one tries to get information on the global structure of the category A-mod. In particular, one is looking for invariants which allow to distinguish the isomorphism classes. Of course, any (finite dimensional) representation can be written as a direct sum of indecomposable representations, and the theorem of Krull-Remak-Schmidt asserts that such a decomposition is unique up to isomorphism. Thus, it usually is sufficient to consider indecomposable representations. Note that in general there will be infinitely many isomorphism classes of indecomposables as already the four dimensional Kronecker algebra shows. In case there are only finitely many isomorphism classes of indecomposable A-modules, the algebra A is said to be representation finite. If we are interested in determining the indecomposable A-modules, we may assume that A is connected: that it has no central idempotents different from 0 and 1, since any non-trivial central idempotent of A yields a product decomposition $A = A_1 \times A_2$, and any indecomposable A-module is either an A_1-module or an A_2-module.

The setting of dealing with a finite dimensional k-algebra was enlarged by Auslander and his school to that of artin algebras: these are rings with center an artinian ring and being module finite over the center. In particular, all finite rings are artin algebras.

Eight Topics in Eight Years

1. The solution of the first Brauer-Thrall conjecture (1968, 1974). The first Brauer-Thrall conjecture asserts that an algebra A is representation finite provided that the indecomposable A-modules are of bounded length. This conjecture was solved in 1968 by Roiter [R] using an interesting ordering of the indecomposables. The method of proof was formalized by Gabriel [G2] by introducing what he calls the Roiter measure of a module M: it is an increasing sequence of numbers and it is given by the lengths of indecomposable submodules M_i occurring in a suitable chain $M_1 \subset M_2 \subset \cdots \subset M_t \subseteq M$. The set of such sequences can be totally ordered and in this way one obtains the ordering used by Roiter (or better: the dual analogue).

Contrary to an assertion in the paper, the proof does not work in the case of an arbitrary artinian ring, since an essential intermediate assertions fails to hold in general. In 1974, Auslander [A2] gave a new proof of the first Brauer-Thrall conjecture, and this proof indeed deals with the general situation of an artinian ring. Auslander's proof can be rewritten as an assertion concerning the Auslander-Reiten quiver of A and one obtains in this way a much stronger assertion: If the Auslander-Reiten quiver of a connected algebra A has a component C which contains only modules of bounded length, then this component is finite and is the only one.

2. The classification of the $k[T_1, T_2]/\langle T_1 T_2 \rangle$-modules (1968). In the same year 1968, a very interesting classification problem was solved by Gelfand and Ponomarev [GP1], namely that of the finite dimensional A-modules, where $A = k[T_1, T_2]/\langle T_1 T_2 \rangle$ (here, $k[T_1, T_2]$ is the usual polynomial ring in two variables T_1, T_2 and $\langle T_1 T_2 \rangle$ is the ideal generated by the element $T_1 T_2$). Of course, the A-modules for this algebra A are just vector spaces endowed with a pair of operators which annihilate each other: the operators are the multiplications by T_1 and by T_2 on the module. Note that A is infinite dimensional, but by adding relations of the form $T_1^n = 0$ and $T_2^m = 0$, we obtain finite dimensional factor algebras $B_{n,m}$ and any indecomposable finite dimensional A-module which is not annihilated by either T_1 or T_2 is actually a $B_{n,m}$-module for some n, m. The problem of classifying the $k[T_1, T_2]/\langle T_1 T_2 \rangle$-modules turned out to be the final step in order to classify certain representations of the Lorentz group, it was in this setting that Gelfand and Ponomarev have attacked it.

The classification principle and the method of proof were later generalized to the socalled special biserial algebras: they are given by a quiver Q and an ideal I of relations with the following two properties: Any vertex of Q is starting point of at most two arrows and also end point of at most two arrows, and for every arrow β there are at most two arrows α, γ such that $\alpha\beta$ and $\beta\gamma$ do not belong to I. These conditions exclude, in the language of the representation theory of quivers to be discussed below, all possible $\mathbb{D}_n$-cases, including the tilted ones. Since A is a factor algebra of the path algebra kQ, we may consider the arrows of Q as elements of A (belonging to $J(A) \setminus J(A)^2$). The description of the indecomposable A-modules is quite easy: It turns out that many of these modules (the string modules and the primitive band modules) come equipped with a k-basis Ξ which can be displayed as a line or a cycle (to be more precise: as the vertices of a quiver of type $\mathbb{A}_n$ or $\tilde{\mathbb{A}}_n$) such that the multiplication of any element of this basis with an arrow α is either zero or a (non-zero) scalar multiple of an element of Ξ (and in the latter case, one has a corresponding arrow in the quiver). We can arrange it in such a way that a scalar different from 1 occurs at most once and only in the case of the cylic quuiver. The information given by this display can be encoded conveniently by a formal word in some alphabet; as alphabet we take two copies of the set of arrows of Q. For example, in the case considered by Gelfand and Ponomarev one takes formal words in the letters $T_1, T_2; T_1^{-1}, T_2^{-1}$. The remaining modules are built from the primitive band modules as iterated self-extensions using (as in the case of $k[T]$-modules) a Jordan normal form.

String modules and band modules can be defined for any k-algebra A.

The decisive property of a special biserial algebra A is the fact that in this case, these are the only indecomposable A-modules. In their proof, Gelfand and Ponomarev had used the theory of additive relations (recall that an endomorphism of a vector space V may be considered as a subspace of $V \times V$, its graph, and additive relations on V are just arbitrary subspaces of $V \times V$), and Gabriel has proposed to rewrite the proof using functorial filtrations of the forgetful functor. Of importance is the fact that two such filtrations are sufficient to work with. Now, given two filtrations of a vector space V, there always is a basis of V which is compatible with both of them; this is the way one obtains a convenient basis as mentioned above.

The description of the A-modules by Gelfand and Ponomarev can be interpreted in terms of the covering theory which later was developed by Gabriel and his students; here we deal with the special situation where all indecomposable representations of the covering are thin (a representation of a quiver is said to be thin provided all vector spaces used are of dimension at most 1).

3. Wildness of algebras (1969). The features of wildness were first exhibited by Corner as occurring for infinite abelian groups. The main idea is the following: for certain rings R, one observes that all kinds of other rings, even those which are considered as very pathological, can be realized as endomorphism rings of R-modules, or at least as factor rings of endomorphism rings of R-modules modulo nice ideals. This then means that the given algebra has modules with (nearly) prescribed endomorphism rings, and, in particular modules with pathological decomposition behaviour. What is of interest for us is the fact that there do exist finite dimensional algebras which are wild in this sense: in 1969, Corner [C] has shown that the 5-subspace problem is wild.

The n-subspace problem asks for the mutual position of n subspaces of a vector space, equivalently, we may consider the following quiver $\Delta(n)$ (called the n-subspace quiver)

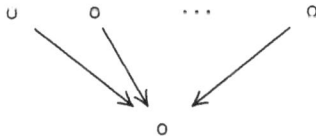

with $n + 1$ vertices and its representations which use only injective maps. Since one easily sees that an indecomposable representation involving maps which are not injective is one dimensional, there is no essential difference between the n-subspace problem and the classification problem for the n-subspace quiver $\Delta(n)$. Here we see very clearly a way for rewriting certain geometrical problems in terms of representations of quivers.

If $n \leq 3$, then it is easy to see that we deal with a representation finite problem: all indecomposable representations of $\Delta(2)$ and all but one indecomposable representations of $\Delta(3)$ are thin; the only exception has dimension vector $\begin{smallmatrix} 1 & & 1 & & 1 \\ & & 2 & & \end{smallmatrix}$ (here, the numbers are displayed in the shape of the quiver and indicate the dimensions of the various vector spaces involved), it is given by the three one dimensional subspaces $k \times 0$, $0 \times k$ and $\{(x, x) \mid x \in k\}$ of $k \times k$.

The case $n = 4$ will be discussed below, let us consider now the case $n = 5$, the five-subspace problem. Let us show why this problem is wild. Namely, consider the following construction: given a vector space V and a pair of endomorphisms ϕ, ψ of V, let $F(V; \phi, \psi)$ be given by considering the vector space $V \times V$ together with the following 5-tuple of subspaces:

$$V \times 0, \quad 0 \times V, \quad \Gamma(1), \quad \Gamma(\phi), \quad \Gamma(\psi),$$

where $\Gamma(-)$ denotes the graph of a given endomorphism: $\Gamma(\phi) = \{(x, \phi(x)) \mid x \in V\}$. Note that a vector space V endowed with a pair of endomorphisms ϕ, ψ of V is nothing else than a $k\langle T_1, T_2 \rangle$-module, where $k\langle T_1, \ldots, T_n \rangle$ is the free associative algebra in n generators, here, V is the underlying vector space of the module, and the multiplication by T_1 is given by the endomorphism ϕ, whereas T_2 acts via ψ. The construction above yields a functor F from the category of $k\langle T_1, T_2 \rangle$-modules to the category of all $k\Delta(5)$-modules, and one can show without difficulty that F is full and faithful, thus a full embedding. This shows, in particular, that any k-algebra which occurs as the endomorphism ring of a $k\langle T_1, T_2 \rangle$-module also occurs as the endomorphism ring of a $k\Delta(5)$-module.

Let us pursue these considerations. Trivially, for any $n \geq 2$, we may embed the category of $k\langle T_1, T_2 \rangle$-modules into the category of $k\langle T_1, \ldots, T_n \rangle$-modules, just consider the $k\langle T_1, \ldots, T_n \rangle$-modules with trivial action by $T_3, \ldots, T_n$. However, the following construction G_n yields a full exact embedding of $k\langle T_1, \ldots, T_n \rangle$-mod into $k\langle T_1, T_2 \rangle$-mod; given a vector space V and n endomorphisms $\phi_1, \ldots, \phi_n$, let $G_n(V; \phi_1, \ldots, \phi_n)$ be defined as

$$\left(\bigoplus_{i=1}^{n+2} V; \begin{bmatrix} 0 & 1 & & & \\ & \ddots & \ddots & & \\ & & \ddots & \ddots & \\ & & & \ddots & 1 \\ & & & & 0 \end{bmatrix}, \begin{bmatrix} 0 & & & & \\ 1 & \ddots & & & \\ \phi_1 & \ddots & \ddots & & \\ & \ddots & \ddots & \ddots & \\ & \phi_n & 1 & 0 \end{bmatrix} \right)$$

(with the remaining entries being zero). Now, if B is any k-algebra which is generated by n elements, then B may be considered as a factor algebra of the k-algebra $k\langle T_1, \ldots, T_n \rangle$, and therefore B-mod occurs as a full exact subcategory of $k\langle T_1, \ldots, T_n \rangle$-mod, say with embedding functor H_B. Altogether, we see that the composition of the functors H_B, G_n and F yields a full exact embedding

$$FG_nH_B: B\text{-mod} \longrightarrow k\Delta(5)\text{-mod}.$$

For example, the path algebra $k\Delta(n)$ of the n-subspace quiver is generated as a k-algebra by $2n$ elements (take n idempotents and the n arrows of the quiver), thus there exists a full exact embedding of the category of all $k\Delta(n)$-modules into the category of all $k\Delta(5)$ modules. As a consequence, the classification problems for the n-subspace problems with $n \geq 5$ may be considered as being essentially not different from each other.

We should add that all these considerations do not have to be restricted to finite dimensional modules, but work in general. Also the k-algebra B does not have to be finitely generated. In the paper [C], Corner had to assume that B has a generating set of cardinality less than the first strongly inaccessible cardinal. Even this (rather weak) condition was later removed by Shelah.

After the work of Corner, many other algebras A have been shown to be wild, see for example [B1,B2]. In particular, the quiver $K(3)$

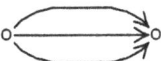

is wild, just consider the full subcategory of all representations of $K(3)$ where the upper arrow is given by an isomorphism.

In general, one should not require that there is a full exact embedding of $k\langle T_1, T_2 \rangle$-mod into A-mod, but a little less, since in A-mod there may be maps which are in some sense inessential, so that it does not matter whether they are in the image of an embedding functor $\iota: k\langle T_1, T_2 \rangle$-mod $\longrightarrow A$-mod or not. Thus, we are going to drop the requirement that ι should be full. Of course, given $k\langle T_1, T_2 \rangle$-modules X and Y, there should be some control concerning the maps in $\operatorname{Hom}_A(\iota X, \iota Y)$ which are not in the image of ι.

For example, consider the algebra $C = k[T_1, T_2, T_3]/I$, where I is generated by all monomials of degree 2. The category B-mod and the category of representations of the quiver $K(3)$ are very similar: If $\mathcal{U}$ is an additive subcategory of A-mod, let $\langle \mathcal{U} \rangle$ denote the ideal of A-mod given by all maps which factor through an object of $\mathcal{U}$, and let $\pi_{\mathcal{U}}$ denote the projection functor A-mod $\longrightarrow A$-mod$/\langle \mathcal{U} \rangle$. Let us denote by $\mathcal{S}$ the subcategory of all semisimple C-modules (since C is a local algebra, the semisimple C-modules are just direct sums of the unique simple C-module). Then the category C-mod$/\langle \mathcal{S} \rangle$ is equivalent to the full subcategory of all representations of

$K(3)$ which do not have a simple direct summand. We can rephrase this as follows: There is an exact embedding functor $\iota \colon K(3)$-mod $\longrightarrow$ B-mod such that for any pair of $K(3)$-modules X, Y the space $\operatorname{Hom}_A(\iota X, \iota Y)$ is the direct sum of $\iota(\operatorname{Hom}_B(X, Y))$ and the space of all homomorphisms $\iota X \to \iota Y$ which belong to $\langle \mathcal{S} \rangle$.

Based on the known examples of what seems to be 'wild' behaviour, an algebra A should be called wild provided there exists an exact functor $k\langle T_1, T_2 \rangle$-mod $\longrightarrow$ A-mod and an additive subcategory $\mathcal{U}$ of A-mod such that the composition of F with $\pi_\mathcal{U}$ is a full embedding. There is a famous tame-and-wild theorem by Drozd, but it uses a slightly different (and more technical) notion of wildness. At the moment, it does not seem to be known whether Drozd's definition of wildness coincides with the one mentioned here.

4. The four-subspace problem (1970). Whereas the five-subspace problem is wild, there does exist a full classification of the indecomposable representations of the four-subspace quiver, as earlier investigations of Nazarova have shown. The (apparently independent) proof by Gelfand and Ponomarev [GP2] in 1970 has been of great importance for the further development since the authors have introduced several new and valuable techniques and they provide remarkable insight into the structure of the module category.

Let A be the path algebra of the four-subspace quiver. Gelfand and Ponomarev introduce what now are called the Coxeter functors C^+ and C^-, these are endofunctors of the category A-mod which allow to construct countably many indecomposable modules starting from the indecomposable injective or the indecomposable projective modules, respectively. For any path algebra of a quiver, it is very easy to construct those indecomposable modules which are projective or injective; the modules obtained from the projectives using powers of C^- are called preprojective, those obtained from the injectives using powers of C^+ are called preinjective. An A-module which does not have any indecomposable preprojective or preinjective direct summand is said to be regular. Surprisingly, the regular modules form an abelian exact subcategory $\mathcal{T}$ of A-mod, and this subcategory is serial in the following sense: Any indecomposable regular module M has a unique chain of regular submodules, thus a unique composition series when considered as an object of the abelian category $\mathcal{T}$. In order to determine all regular indecomposable modules, it is sufficient to construct those which are simple as objects in $\mathcal{T}$, and then to form suitable extensions. It turns out that there are six isomorphism classes of simple objects in $\mathcal{T}$ which behave in a special way: these are the thin representations with dimension vectors of the form $\begin{smallmatrix} d_1 & d_2 & d_3 & d_4 \\ & 1 & \end{smallmatrix}$, where two of the numbers d_i are equal to 1 and the remainig two are equal to 0. The remaining simple objects in $\mathcal{T}$ have (if k is algebraically closed) dimension vector $\begin{smallmatrix} & 1 & 1 & 1 & 1 \\ & & 2 & & \end{smallmatrix}$ and are given by four-tuples of pairwise different one dimensional subspaces in a vector space of dimension 2. To deal with four-tuples of one dimensional subspaces in a vector space of dimension 2 means to consider four points on a projective line, and there is the well-known cross ratio which occurs as an invariant with respect to projective equivalence. The classification of all possible representations of the four-subspace quiver allows a complete understanding of the mutual position of four subspaces in any projective space. Gelfand and Ponomarev also provide a numerical criterion which allows to decide whether a given indecompable A-module M is preprojective, regular or preinjective, just knowing the dimension vector $\mathbf{d} = \begin{smallmatrix} d_1 & d_2 & d_3 & d_4 \\ & d_0 & & \end{smallmatrix}$ of M. By definition, the defect of M (or of $\mathbf{d}$) is

$$\delta(M) = \delta(\mathbf{d}) = 2d_0 - \sum_{i=1}^{4} d_i,$$

and $\delta(M)$ is negative, or zero, or positive if and only if M is preprojective, or regular, or preinjective, respectively. Note that the defect compares the sum of the dimensions of the four subspaces with twice the dimension of the total space. This numerical distinction has its basis in the global structure of the module category: the regular representations form what now is called a separating tubular family, they separate the preprojective modules from the preinjective ones.

The four-subspace quiver is a special 'tame' quiver, the remaining tame quivers will be mentioned below: as we will see they have very similar properties. One particular application of the representation theory of the n-subspace quivers should be mentioned. Let us denote by $L(n)$ the free modular lattice in n generators. The lattice $L(3)$ is finite, it was considered by Dedekind in 1900. The lattices $L(n)$, for $n \geq 4$, are infinite and Gelfand and Ponomarev have used the preprojective and the preinjective representations of $\Delta(n)$, in order to shed light on the structure of $L(n)$.

5. Auslander algebras (1971). In his famous Queen Mary Notes [A1], M. Auslander has given an interesting characterization of the representation finite algebras. Let A be representation finite, and let $M_1, \ldots, M_n$ be indecomposable A-modules, at least one from each isomorphism class. Then $M = \bigoplus_{i=1}^{n} M_i$ is an additive generator of A-mod (this means that any A-module is a direct summand of some power M^m). Let E be the endomorphism ring of M, this is again a finite dimensional k-algebra and it has the following two properties: its global dimension is at most 2, its dominant dimension is at least 2; we recall that to assert that the dominant dimension is at least t means that the terms $I_0, \ldots, I_{t-1}$ in the minimal injective resolution

$$0 \to {}_A A \to I_0 \to I_1 \to \cdots \to I_n \to \cdots$$

of ${}_A A$ are projective. Also conversely, every algebra B of global dimension at most 2 and dominant dimension at least 2 is obtained in this way, and one calls such an algebra nowadays an Auslander algebra. Actually, the correspondence between A and E defines a bijection between the Morita equivalence classes of the representation finite algebras A and the Morita equivalence classes of Auslander algebras. To recover an algebra in the Morita equivalence class of A from the Auslander algebra E, let N be an E-module which is both projective and injective and which contains as a direct summand any indecomposable E-module which is both projective and injective; then the endomorphism ring of N is as required.

If we start with a basic representation finite algebra A and M is an additive generator of A-mod, then almost always the endomorphism ring of M will be larger than A (the only exception occurs in the uninteresting case when A is semisimple). One may wonder whether it is of interest for the study of representation finite algebras to deal with the much larger Auslander algebras. However, it turns out that the Auslander algebras are much better behaved and easier to visualize. So, when Riedtmann started to classify the self-injective representation finite algebras, she looked at the corresponding Auslander algebras and classified them. Of course, in the terminology to be introduced below, for k an algebraically closed field of characteristic different from 2, the basic Auslander algebras are just the path algebras of the Auslander-Reiten quivers of representation finite algebras modulo the mesh relations.

Auslander's result has an interesting consequence: if A is representation finite, then any not necessarily finite dimensional module is a direct sum of copies of finite dimensional indecomposable modules. The reason is the following: let M be an additive generator of A-mod, and E its endomorphism ring. Then $\operatorname{Hom}_A(M, -)$ defines an embedding of the category of all A-modules into the category of all E-modules, and the image is just the full subcategory of all projective E-modules. But every projective E-module is a direct sum of the finite dimensional indecomposable projective E-modules, and these are the images of the finite dimensional indecomposable A-modules.

The algebras of dominant dimension at least two are, in the terminology used in [T], the QF-3 maximal quotient algebras. Let us recall the relevant arguments: As defined by Thrall in 1948, a finite dimensional algebra A is said to be QF-3 provided it has a minimal faithful module M

(this means that M is faithful, and that M occurs as a direct summand of any faithful A-module). Clearly, such a minimal faithful module has to be both projective and injective. It follows that A is a QF-3 algebra if and only if there exists a faithful module which is both projective and injective, thus if and only if the injective envelope of ${}_A A$ is projective. Thus we see that the QF-3 algebras are just the algebras of dominant dimension at least 1. Let $I_0 = I({}_A A)$ be the injective envelope of ${}_A A$, and let B be the maximal submodule of I_0 containing ${}_A A$ such that no composition factor of $\overline{B} = B/{}_A A$ is embeddable into ${}_A A$. This implies that we have both $\text{Hom}_A(\overline{B}, B) = 0$ and $\text{Ext}^1_A(\overline{B}, B) = 0$. If we apply $\text{Hom}_A(-, B)$ to the exact sequence $0 \to {}_A A \to B \to \overline{B} \to 0$, we see that the inclusion map $\iota: {}_A A \to B$ induces an isomorphism $\iota^*: \text{Hom}_A(B, B) \to \text{Hom}_A({}_A A, B)$. The last set can be identified with the underlying set of B, thus ι^* shows that this set B carries the structure of an algebra (namely the endomorphism algebra of ${}_A B$), and clearly the algebra A is embedded into B as a subalgebra. One calls B the classical quotient algebra of A, and A is said to be a maximal quotient algebra provided $A = B$. Of course, one has $A = B$ if and only if all the composition factors of the socle of $I_0/{}_A A$ occur in the socle of ${}_A A$, thus if and only if the injective envelope I_1 of $I_0/{}_A A$ belongs to the additive subcategory generated by I_0.

6. Representations of quivers (1972, 1973, 1974).

At the same time, when Auslander presented this general characterization of representation finite algebras, Gabriel [G1] was considering in detail the representation finite hereditary k-algebras A, where k is an algebraically closed field. So let us now assume that A is a connected hereditary k-algebra and that k is an algebraically closed field. Up to Morita equivalence, $A = kQ$ where Q is a finite quiver without oriented cycles. What Gabriel has shown is that A is representation finite if and only if Q is of the form $\mathbb{A}_n, \mathbb{D}_n, \mathbb{E}_6, \mathbb{E}_7$ or $\mathbb{E}_8$ (this means that after forgetting the orientation of the arrows, we obtain from the quiver one of the simply laced Dynkin diagrams which occur in the structure theory of semisimple complex Lie algebras):

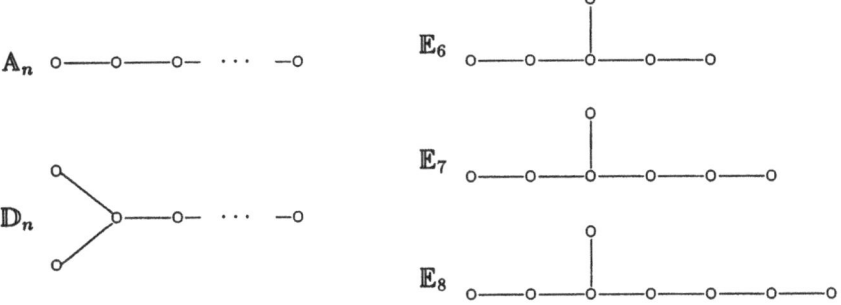

Moreover, in these cases, the dimension vectors of the indecomposable representations are just the positive roots of the corresponding quadratic form occurring in Lie theory (as the Killing form) and the dimension vector determines such an indecomposable representation up to isomorphism. Thus, the dimension vectors are convenient invariants in order to distinguish isomorphism classes.

Gabriel's work was meant as a correction and completion of earlier (unreadable and erroneous) investigations of Yoshii, and some other authors (Krugliak, Bäckström) have presented corresponding corrections at the same time. However, it was the graphical interpretation, the translation to combinatorics and the observation that there is a surprising relationship to Lie theory which marks the value of this work and which has stimulated the further research. Indeed, already one year later (1973) Bernstein, Gelfand and Ponomarev [BGP] gave a new proof concentrating on the relationship to Lie theory. Part of their work deals with the representation theory of an arbitrary finite quiver without oriented cycles, and they use in a systematic way Coxeter functors in order to construct the corresponding preprojective and preinjective representations. The Dynkin cases $\mathbb{A}_n, \mathbb{D}_n, \mathbb{E}_6, \mathbb{E}_7$ or $\mathbb{E}_8$ are characterized as the only cases with only finitely many indecomposable preprojectives; this condition alone already implies that all the indecomposables are preprojective.

Let us come back to the four-subspace quiver, it is a quiver of type $\tilde{\mathbb{D}}_4$, and it turns out that similar results hold true for all quivers whose type is one of the extended Dynkin diagrams $\tilde{\mathbb{A}}_n, \tilde{\mathbb{D}}_n, \tilde{\mathbb{E}}_6, \tilde{\mathbb{E}}_7$ or $\tilde{\mathbb{E}}_8$. These quivers are said to be the tame ones, and they have been investigated by Donovan-Freislich [DF] and by Nazarova [N]. Again, the indecomposables are of three different kinds: there are the preprojectives, the preinjectives and the regular ones, and there is again the notion of a defect which yields a numerical criterion in order to distinguish these cases. For the distinction between tame and wild quivers we refer to Brenner (1974) [B2].

Our joint work with Dlab [DR1, DR2] has extended these investigations to the case of an arbitrary, not necessarily algebraically closed base field k, dealing with species (as introduced by Gabriel [G2]) instead of quivers. It is interesting that in this way all possible Dynkin diagrams $\mathbb{A}_n, \mathbb{B}_n, \ldots, \mathbb{G}_2$ (and not only the simply laced ones) appear when dealing with a representation-finite hereditary algebra; similarly all the extended Dynkin diagrams $\tilde{\mathbb{A}}_n = \mathbb{A}_n^{(1)}, \ldots, \mathbb{G}_2^{(1)}, \ldots, \mathbb{D}_4^{(3)}$ appear in the tame case. A diagram such as $\mathbb{B}_2$ or $\mathbb{G}_2$ encodes a skewfield embedding $F \subset G$ of degree 2 or 3. A basic algebra of type $\mathbb{B}_2$ or $\mathbb{G}_2$ is of the form $\begin{bmatrix} F & G \\ 0 & G \end{bmatrix}$ or $\begin{bmatrix} G & G \\ 0 & F \end{bmatrix}$.

The relationship to Lie theory has further been strengthened in the meantime. First of all, Kac has shown that the dimension vectors of the indecomposable representations of any finite quiver are just the corresponding positive roots. For any positive real root, there is a unique isomorphism class of indecomposable representations, whereas for any positive imaginary root, there is a proper family of such isomorphism classes. Second, one can use the representation theory of quivers and species in order to construct the complex semisimple Lie algebras (or Kac-Moody algebras) $\mathbf{g}$ and even their Chevalley $\mathbb{Z}$-forms. Let us concentrate on the positive part $\mathbf{n}_+$ of $\mathbf{g}$. The quantizations of the Serre relations for $\mathbf{n}_+$ as introduced by Drinfel'd and Jimbo express universal relations between the possible composition series of representations; in three different ways quiver varieties have been used: working over finite fields, one may count composition series, one may use perverse sheaves working over an l-adic field, or else methods from differential geometry when working over $\mathbb{R}$.

Also, the reflection functors used to construct the Coxeter functor have attracted a lot of interest. After all, they allow to compare the representations of two algebras which are not Morita equivalent. The tilting functors were introduced as a broad generalization of these reflection functors.

7. Representations of posets (1972, 1975). One of the main tools in the representation theory of algebras is the use of representations of posets and of S-spaces, where S is a poset. A systematic study was started by Nazarova and Roiter [NR]. They and their students developed, on the one hand, techniques for studying representations of posets, and, on the other hand, they showed in which way many problems in algebra can be reduced to problems dealing with representations of posets. As in the case of the hereditary algebras, it was the use of graphical methods, of quadratic forms and root systems which attracted the interest and stimulated the further research. In particular, two results of Kleiner were found to be very useful: the classification of the minimal representation infinite posets [K1] and of the sincere representation finite posets [K2]. Here is the list of the minimal representation infinite posets:

$$(1,1,1,1) \qquad (2,2,2) \qquad (3,3,1) \qquad (4,N) \qquad (5,2,1) \, ,$$

where $(c_1, \ldots, c_n)$ stands for the disjoint union of n chains with $c_1, \ldots, c_n$ elements, respectively, and $(4, N)$ is the disjoint union of a chain of 4 elements and the poset N. A functorial way of constructing indecomposable representations of a finite poset, similar to the use of Coxeter functors for quivers, was designed by Drozd [D] and he showed that quadratic forms and their roots play a similar role as for quivers. In [N2], Nazarova (1975) was able to characterize the posets which are of wild type: Here is the list of the minimal wild posets:

$$(1,1,1,1,1) \qquad (2,1,1,1) \qquad (3,2,2) \qquad (4,3,1) \qquad (5,N) \qquad (6,2,1).$$

We did not yet define what is meant by the representation theory of posets. Actually, there are two competing ways: The first one, introduced by Nazarova and Roiter, deals with matrix representations. The aim is to bring a given matrix whose columns are labelled by the elements of the given poset S into normal form: here, arbitrary row transformation, but only suitable column transformations (depending on the ordering of S) are allowed. The second one, that of an S-space, was popularized by Gabriel [G3]: here, one deals with a vector space V_ω and subspaces $V_s \subseteq V_\omega$, where s is an element of S and one requires that for $s \le t$, one has $V_s \subseteq V_t$. For example, the n-subspace problem just deals with the classification of all S-spaces where S is the discrete poset with n elements, whereas the consideration of flags in a vector space means that one deals with S-spaces where S is a chain. It is worthwhile to note that the two theories are quite similar: The category $\mathrm{rep}(S)$ of matrix representations of S contains trivial objects 0_s indexed by the elements of S, and if we factor out from $\mathrm{rep}(S)$ the ideal of all maps which factor through direct sums of copies of these objects 0_s, then we obtain a category which is equivalent to the category of all S-spaces.

We may consider the category of S-spaces as a full subcategory of a module category: Let S^+ be obtained from S by adding an element ω with $s < \omega$ for all $s \in S$. Let $I(S^+)$ be the incidence algebra of S^+; its quiver is given by the Hasse diagram of S^+ and as relations one has to take all commutativity relations. Obviously, any S-space $(V_\omega; V_s)$ may be considered as a representation of this quiver using as maps the inclusion maps, thus also the commutativity relations are satisfied. Note that the incidence algebras $I(S^+)$ of the posets $(1,1,1,1), (2,2,2), (3,3,1)$ and $(5,4,1)$ are just hereditary algebras of type $\tilde{\mathbb{D}}_4, \tilde{\mathbb{E}}_6, \tilde{\mathbb{E}}_7$ and $\tilde{\mathbb{E}}_8$, for $S = (4, N)$ we obtain a tilted algebra of type $\tilde{\mathbb{E}}_8$; always, all but finitely many indecomposable $I(S^+)$-modules are actually obtained from S-spaces. Thus, we see that there is a strong relationship between the representation theory of posets and of quivers.

When Gabriel dealt with the representation finite quivers, he used this relationship and worked with S-spaces. There always was the feeling that the classification problem for modules over a k-algebra A, where k is an algebraically closed field, should be reduceable to the handling of suitable posets. In case A is representation finite, one may fix an indecomposable projective A-module P, and one can show that the hammock given by the functor $\mathrm{Hom}_A(P, -)$ corresponds to the category of S-spaces for a (uniquely defined) representation finite poset S, thus subspace considerations are sufficient in order to deal at least with representation finite algebras.

8. The existence of almost split sequences (1975).
In 1974, Auslander began to publish a series of papers entitled *Representation theory of*

artin algebras. The title already indicates that this series was meant to lay the foundation of a general theory. The first two parts headed for a proof of the first Brauer-Thrall conjecture and included a lot of additional basic material. Starting with the third part [AR1], Reiten became a coauthor. It is this third part which had the greatest impact as it contains the concept of and the first existence proof for the almost split sequences.

The existence of almost split sequences is derived by specializing one of the numerous compatibility formulae from Cartan-Eilenberg: given rings A, R, a left A-module M, an R-A-bimodule M' and an injective left R-module I, there is a canonical isomorphism

$$\rho \colon \operatorname{Ext}_A^1(M, \operatorname{Hom}_R(M', I)) \longrightarrow \operatorname{Hom}_R(\operatorname{Tor}_1^A(M', M), I).$$

Auslander and Reiten consider an artin algebra A with center R and take as I the minimal injective cogenerator in R-mod; for any R-module X, we denote by $DX = \operatorname{Hom}_R(X, I)$ its dual. The A-module M is arbitrary (but of finite length), and M' is its transpose: Take a minimal projective presentation of M, say write M as the cokernel of a map $p \colon P_1 \to P_0$ where P_0, P_1 are projective and of minimal length, then the transpose $\operatorname{Tr} M$ of M is just the cokernel of $\operatorname{Hom}_A(p, {}_A A)$. Note that the transpose of a (left) A-module M is always a right A-module. Of course, starting with a right A-module M', we obtain as DM' a left A-module. Also, Auslander and Reiten observe that $\operatorname{Tor}_1^A(\operatorname{Tr} M, M)$ is nothing else but the factor $\underline{\operatorname{End}}_A(M)$ of the endomorphism ring $\operatorname{End}_A(M)$ modulo the subgroup $P(M)$ of those endomorphisms of M which factor through a projective module, thus the right hand side is $D \underline{\operatorname{End}}_A(M)$, and we obtain an isomorphism

$$\rho \colon \operatorname{Ext}_A^1(M, D \operatorname{Tr} M) \longrightarrow D \underline{\operatorname{End}}_A(M).$$

Assume now that M is indecomposable and not projective. Then $\operatorname{End}_A(M)$ is a local ring and $\underline{\operatorname{End}}_A(M)$ is a non-zero factor ring. The elements of $D \underline{\operatorname{End}}_A(M)$ are those R-linear maps $\alpha \colon \operatorname{End}_A(M) \to I$ which vanish on $P(M)$. If α is such a map, non-zero and vanishing on the radical of $\operatorname{End}_A(M)$, then $\rho^{-1}(\alpha)$ yields an exact sequence with rather peculiar properties: it is non-split, but it is almost split: any exact sequence which is non-trivially induced from it will split.

The importance of these almost split sequences cannot be overestimated, they are now one of the main general tools in representation theory. There are many different ways to interpret them. Both maps occurring in an almost split sequence, the non-split monomorphism and the non-split epimorphism are irreducible maps: they do not have non-trivial factorizations. These maps provide a two-step connection between the module M and the dual $D \operatorname{Tr} M$ of its transpose. Actually, the construction $D \operatorname{Tr}$ which yields a bijection between the indecomposable non-projective modules and the indecomposable non-injective modules is a major tool for constructing new indecomposable modules from given one. In some sense, one may say that the almost split sequences form a rather rigid corset of the module category. This shall be explained below in more detail.

Parallel to this series of papers, Auslander and Reiten published a second series of papers [AR2] with a rather technical name: *Stable equivalence of dualizing R-varieties*. Here, the considerations are developed in a much wider context, starting with functor categories and showing that certain properties of additive categories are preserved when one replaces the additive category $\mathcal{A}$ by the category mod($\mathcal{A}$) of finitely presented functors. Of importance is the fact that the simple functors are finitely presented, and a minimal presentation of such a simple functor just yields an almost split sequence. In this way, the existence of almost split sequences has to be considered as a rather surprizing finiteness property.

Remark concerning the contacts between the different centers. At the beginning of the period which we report on, the various research groups were working quite independently. At the Oberwolfach meeting on commutative algebra in 1971, Auslander and Gabriel reported their results concerning (non-commutative) representation finite algebras [A1,G1], and they discussed the importance of further investigations. The survey [G2] by Gabriel was decisive in spreading knowledge about the new developments. The Bonn workshop 1973 brought together some of the specialists working in the field. At that time, Dlab took the initiative to start a series of meetings under the heading *International Congress on Representations of Algebras*, the first such ICRA was organized at Carleton University Ottawa August 1975 and was very successful. The publication of the Ottawa proceedings in 1975 by Dlab and Gabriel (Springer LNM 488) served as a first public forum for the presentation of the different approaches.

The Representation Types

A glance at the eight topics mentioned above shows that almost all are centered around the possible representation type of a given algebra. Indeed, only the last topic is directed towards the understanding of the representations of a general finite dimensional algebra. Let us discuss here the first seven topics from a systematic point of view, the evaluation of the last topic will be postponed to the next section.

The examples collected during the period which we discuss suggested that there should be a division into three different representation types: finite, tame and wild. Let us recall that the representation finite algebras are those which have up to isomorphism just a finite number of indecomposable modules. The remaining ones, the representation infinite algebras, may have the wild behaviour as described above, and in this case it should be hopeless to find a complete description of all isomorphism classes, at least there are families of pairwise non-isomorphic indecomposable modules which are parametrized by an arbitrary number of parameters. On the other hand, there are representation infinite algebras which are 'tame': a complete classification is known, and irreducible families of indecomposables are parametrizable by a single parameter. The conjecture that there is such a trichotomy was formulated during the Bonn workshop in 1973 by Donovan and Freislich, and there were furious discussions about their suggestion that the relevant parameter spaces may be rational: that the one parameter families of a tame algebra may be indexed by a line, and that the wild algebras should have families of indecomposables indexed by linear spaces of arbitrarily high dimension. At that time, the proposed definitions were rather vague. The definitions for tame and wild which are in use now are due to Drozd, using Bocs reduction he was able to prove that there really is such a trichotomy. The investigations of Drozd also confirm the rationality conjecture of Donovan and Freislich, and a stronger version has been established by Gabriel, Nazarova, Roiter, Sergejchuk and Vossieck.

Of particular interest is a theorem of Crawley-Boevey which asserts the following: If A is a tame algebra, then for every natural number d, almost all indecomposable modules of dimension d belong to homogeneous tubes. The modules which belong to a fixed homogeneous tube behave similar to the Jordan blocks $J(\lambda, n)$ with a fixed eigenvalue λ (recall that the Jordan blocks classify the indecomposable finite dimensional $k[T]$-modules, and those for a fixed eigenvalue λ form a homogeneous tube). A homogeneous tube T always contains an indecomposable module M_1 of smallest length, and the indecomposables in T can be arranged as a sequence

$$M_1 \subset M_2 \subset \cdots \subset M_n \subset M_{n+1} \subset \cdots,$$

with M_m/M_n isomorphic to M_{m-n} for all $m > n \geq 1$. In particular, the modules in T are iterated extensions of copies of M_1. The modules of the form M_1 are said to be primitive, and one is interested to determine all possible primitive modules.

The first case which should be considered are the tame algebras whose primitive modules are of bounded length, so that there is only a finite number of one-parameter families of primitive modules. These algebras are said to be domestic. They are the representation infinite algebras which are closest to the representation finite ones. Typical examples are the tame hereditary algebras. Whereas there does exist a reasonable structure theory for the representation finite algebras and their indecomposable representations, not much is known yet about the domestic algebras in general.

Now we come to the non-domestic tame algebras. The only example we have seen above were the special biserial algebras studied by Gelfand and Ponomarev, namely the algebras $B_{n,m} = k[T_1, T_2]/\langle T_1 T_2, T_1^n, T_2^m \rangle$, where $n \geq m \geq 2$ and $n > 2$. The case $n = m = 2$ has to be excluded, since the algebra $B_{2,2}$ is domestic: all the primitive modules are of dimension 2 and there is a unique one-parameter family of such modules. Let us consider now the algebra $B_{3,2}$. One observes that for any natural number e, there are at least 2^e disjoint one-parameter families of primitive modules of dimension $6e + 5$. (For any sequence $\varepsilon = (\varepsilon_1, \ldots, \varepsilon_e)$ of zeros and ones, we construct the following word $T_1 T_2^{-1} T_1^2 T_2^{-1} w(\varepsilon_1) \cdots w(\varepsilon_e)$ where $w(0) = (T_1 T_2^{-1})^3$ and $w(1) = (T_1^2 T_2^{-1})^2$; these kinds of words give rise to primitive band modules and one checks without difficulties that we obtain in this way non-isomorphic ones.) It follows that the number of disjoint one-parameter families of primitive modules of dimension d cannot be bounded by a polynomial $f(d)$: one says that we deal with non-polynomial growth.

There do exist tame algebras which are non-domestic but where the number of one-parameter families of primitive modules is at most of polynomial growth (in the known examples we really have at most linear growth): the tubular algebras, but they were not yet known at that time.

Looking at the number of one-parameter families of primitive modules, one observes a hierarchy: domestic, polynomial growth, non-polynomial growth. However, the known examples suggest that the non-polynomial growth behaviour is paired with a special finiteness condition: the finite tubular type. In order to define this, we need the notion of a generic module as introduced by Crawley-Boevey.

An indecomposable A-module M of infinite dimension is said to be generic provided M considered as an $\text{End}_A(M)$-module is of finite length. A typical example of a generic module is the following representation of the Kronecker quiver

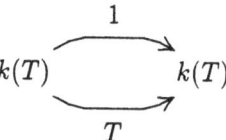

here, $k(T)$ is the field of rational functions in one variable T over the base field k, and the maps are multiplication maps as indicated.

Let A be a tame k-algebra over an algebraically closed field k. An indecomposable A-module P will be said to be a Prüfer module provided there exists a locally nilpotent surjective endomorphism of P with kernel indecomposable and of finite length, such that the following additional condition is satisfied: if U is a submodule of P of finite length, and $P/U = P_1 \oplus P_2$, then one of the modules P_1, P_2 is finite dimensional. Such a Prüfer module has to be infinite dimensional, since a surjective endomorphism of a finite dimensional vector space V can only be locally nilpotent in case $V = 0$. Given a generic A-module M, let $\mathcal{S}(M)$ be the set of isomorphism classes $[X]$ of Prüfer modules X which are direct summands of modules of the form M/M', where M' is a finitely generated A-submodule of M. Call two elements $[X], [Y]$ of $\mathcal{S}(M)$ equivalent provided there are epimorphisms $X \to Y$ and $Y \to X$ (actually, it seems that for modules $X, Y \in \mathcal{S}(M)$ there is an epimorphism $X \to Y$ if and only if there is an epimorphism $Y \to X$). The tubular type of M is given by the cardinalities of the various equivalence classes. We conjecture that all these equivalence classes are finite and that almost all consist of a single element. If there are n equivalence classes of cardinality $r_1 \geq r_2 \geq \cdots \geq r_n \geq 2$, whereas all other equivalence classes are singletons, then the tubular type of M is said to be $r(M) = (r_1, \ldots, r_n)$. Now finite tubular type should mean that either $n \leq 2$ or else $n = 3$ and $r(M)$ is of the form $(r, 2, 2)$ with $r \geq 2$ or of the form $(s, 3, 2)$ with $3 \leq s \leq 5$. Instead of looking at these n-tuples $r(M) = (r_1, \ldots, r_n)$, we may draw a star $\mathbb{T}_M = \mathbb{T}_{r(M)}$ with n arms of the appropriate sizes; it is obtained from the diagrams $\mathbb{A}_{r_1}, \ldots, \mathbb{A}_{r_n}$ by identifying one endpoint each. In this way, $(r, 2, 2)$ yields the diagram $\mathbb{D}_{r+2}$, and one obtains from $(s, 3, 2)$ the diagram $\mathbb{E}_{s+3}$. Thus, to say that M has finite tubular type just means that the star $\mathbb{T}_M$ is of the form $\mathbb{A}_n, \mathbb{D}_n, \mathbb{E}_6, \mathbb{E}_7$, of $\mathbb{E}_8$.

For a tubular algebra A, all but two of the generic modules have tubular type $(2, 2, 2, 2)$, $(3, 3, 3)$, $(4, 4, 2)$ or $(6, 3, 2)$, thus tubular algebras are not of finite tubular type. On the other hand, all the tubular types which appear for the generic modules of a non-domestic special biserial algebra should be of the form $\mathbb{A}_n$. The socalled Gelfand problem deals with tame algebras which are not polynomial growth algebras and here we find generic modules of tubular type $\mathbb{D}_n$ for various n. It seems that the non-polynomial growth behaviour only occurs when the tubular types of the generic modules involved are of the form $\mathbb{A}_n$ or $\mathbb{D}_n$.

Altogether, we expect that there is a hierarchy of representation types as shown in the following table; here, going down means an increase of complexity:

finite

domestic	

polynomial growth	finite tubular type

wild

The three middle boxes are the tame types. Of course, the module category of an individual algebra may have parts which are of finite type whereas other parts are of wild type, so some mixture will occur. But one should be able to separate rather well the different parts.

We have mentioned above that the first seven topics discussed in our survey are dealing with algebras of various representation types. Thus, let us end this section by inserting the corresponding section numbers at the appropriate slots:

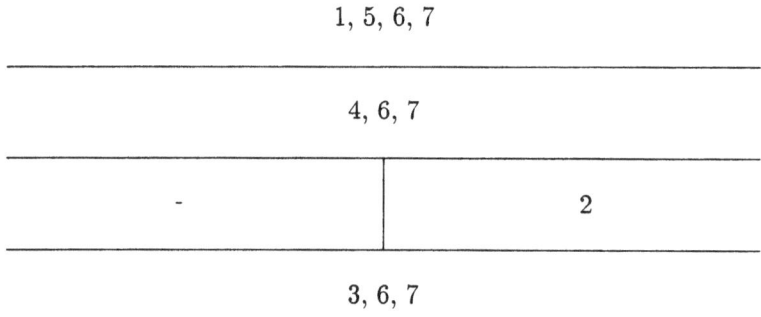

$$1, 5, 6, 7$$

$$4, 6, 7$$

| - | 2 |

$$3, 6, 7$$

The Basic Concepts

It remains to present a small outline in which way the almost split sequences are used as a basic framework. The information given by these sequences is nowadays encoded in the Auslander-Reiten quiver of the algebra.

Thus, let now A denote an artin algebra. Let X, Y be A-modules. A homomorphism $f \colon X \to Y$ is said to be irreducible provided f is neither a split epimorphism, nor a split monomorphism, but for every factorization

$$X \xrightarrow{f_1} I \xrightarrow{f_2} Y$$

of f, the map f_1 is a split monomorphism or f_2 is a split epimorphism. If we consider an additive category as something like a ring, then irreducible homomorphisms are just the 'irreducible' or 'prime' elements: those elements which have only trivial factorizations.

Auslander and Reiten have shown: For every indecomposable A-module X, there exists a map $f \colon X \to Y$ with the following properties: first, any $h \colon X \to Z$ which is not a split monomorphism, can be factored through f, and second, if $Y = Y' \oplus Y''$ with $f(X) \subseteq Y'$, then $Y'' = 0$. Such a map f is (up to ismorphism) uniquely determined by X. One calls $f \colon X \to Y$ the minimal left almost split map starting in X (or the 'source' map for X).

There is also the dual assertion: For every indecomposable module Z there exists a minimal right almost split map $g\colon Y \to Z$ (a 'sink' map).

These source maps and sink maps usually combine to form the almost split sequences: Let X be an indecomposable module which is not injective. Let $f\colon X \to Y$ be the source map starting in X. Then f is injective and its cokernel $Y \to Z$ is the sink map for Z. The module Z is indecomposable and not projective, and every indecomposable non-projective module is obtained in this way.

Of particular importance is the following fact: the sink maps and the source maps are always irreducible maps, and they allow to obtain all irreducible maps between indecomposable modules, as follows: Consider a fixed indecomposable module X, and let $f\colon X \to Y$ be the source map for X. If $p\colon Y \to Y'$ is the projection of Y onto an indecomposable direct summand of Y, then pf is an irreducible map starting in X and ending in an indecomposable module, and all such irreducible maps are obtained in this way. Using also the dual statement, we see that an irreducible map $f : M \to N$ with M, N indecomposable A-modules can be seen in two different ways: as part of the source map for M or else as part of the sink map for N.

The existence of almost split sequences shows that there are many irreducible maps in A-mod and that one obtains at least partial factorizations of given maps involving irreducible maps. As in the case of any ring, one is interested in such factorizations into 'prime' elements. But we have to be quite careful. First of all, we usually will not get factorizations, but finite sums of maps which have factorizations. Also, the process of factorization may not stop, just consider a hereditary algebra of infinite representation type, and a map from a projective module to a regular or an injective module. Finally, such factorizations will not be unique, as any Auslander-Reiten sequence with two middle terms shows:

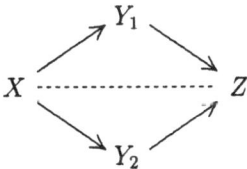

The Auslander-Reiten quiver of A is defined as follows: Its vertices are the isomorphism classes of the indecomposable modules, and there is an arrow $[X] \to [Y]$ provided there exists an irreducible map $X \to Y$. In addition, one usually marks the Auslander-Reiten translation using a dashed line between $[X]$ and $[Z]$ provided there exists an almost split sequence $0 \to X \to Y \to Z \to 0$.

Given an almost split sequence $X \to Y \to Z$, decompose $Y = \bigoplus_{i=1}^{t} Y_i$, with Y_i indecomposable. We get irreducible maps $X \to Y_i$ and $Y_i \to Z$ and

therefore the following meshes

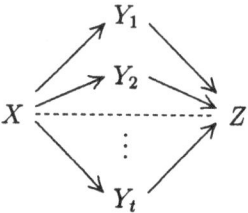

The structure of the Auslander-Reiten quiver of A serves as an important source of information concerning A. The existence of special kinds of components will indicate the special nature of an algebra. Of particular interest are preprojective and preinjective components, since they allow to recover at least part of the algebra immediately. In the previous section, we have seen the importance of tubes, especially homogeneous tubes. Many present investigations are directed towards an understanding of the possible structure of the components and the corresponding hidden information.

Seven Plus One?

One now may be tempted to reformulate the first main heading *Eight topics ...* as *Seven topics plus one ...* . As we have seen, the two years 1968 and 1975 have to be considered as turning-points in the development of the representation theory. So what about the years in between? Our report above seems to indicate a break between the first seven topics which are joint by an obvious thematic unity and the last one.

However, the year 1975 and the appearance of the almost split sequences cannot be seen (and never has been seen) isolated: it is not a coincidence that this concept appears in part III of a longer series. Of course, when Auslander started to conceive part I and part II, he did not know that there should exist almost split sequences in general. Indeed, he once mentioned that in his first thoughts he had tried to correlate the existence of almost split sequences as a strong finiteness condition to the representation finiteness of algebras. It was clear for him for a long time that there are almost split sequences for representation finite algebras, so there was the question whether finite type was a necessary condition or whether almost split sequences existed more generally. As it turned out, and as it is shown in part III, any finite dimensional algebra has almost split sequences. It should be observed that the considerations of part III and the further parts fit naturally into the context of the series which started with parts I and II.

Let us stress in which way the Auslander-Reiten work dealing with almost split sequences is rooted in the previous investigations, showing in this way the natural connection of the eight topics considered. First of all, the first almost split sequences were encountered in the representation finite case. Namely, if we want to present an Auslander algebra by quivers and relations, we have to use the irreducible maps between indecomposable modules in order to obtain the arrows and the almost split sequences as relations. The global dimension 2 of an Auslander algebra E is intimately connected with the existence of almost split sequences: as we have mentioned, they provide minimal projective resolutions of the simple E-modules.

We return to the year 1968 as the year when the solution of the first Brauer-Thrall conjecture was published. The development of the Auslander-Reiten theory is strongly linked to this topic. The Auslander-Reiten theory is nowadays considered as the natural basis for dealing with the first Brauer-Thrall conjecture. Also Roiter's ordering of the indecomposable representations was simulated by Auslander and Smalø in terms of the Auslander-Reiten theory, when the preprojective partition was introduced.

The Auslander-Reiten theory is centered around the construction $D\,\mathrm{Tr}$, that of the dual of the transpose of a module. Note that this construction has a similar effect and may be used in the same way as the Coxeter functors. The similarity was first analyzed by Brenner and Butler, and Gabriel showed the precise relationship: consider the path algebra of a finite quiver without oriented cycles, so that both constructions are defined. We are dealing with two endofunctors which just differ by a sign σ. If V is a representation of the quiver Q, say given by the vector spaces V_i (where i is a vertex of Q) and by linear maps $V_\alpha\colon V_i \to V_j$ (where $\alpha\colon i \to j$ is an arrow) we define $\sigma V = (V_i; -V_\alpha)$. Also, given two representations $V = (V_i; V_\alpha), W = (W_i; W_\alpha)$ of Q, a map $f = (f_i)\colon V \to W$ is given by maps $f_i\colon V_i \to W_i$ satisfying the obvious commutativity relations, and we define $\sigma f = (f_i)_i$. We obtain in this way a functor $\sigma\colon kQ\text{-mod} \longrightarrow kQ\text{-mod}$ which is an automorphism of the category. What Gabriel has shown is that we may identify the Auslander-Reiten functor $D\,\mathrm{Tr}$ with σC^+, but not necessarily with C^+ itself. In order to see the difference, we note the following: If the underlying graph of Q is a tree or, more generally, does not have cycles of odd length, then σV is isomorphic to V, for every object V. But this is no longer true if there are odd cycles and the characteristic of k is different from 2. For example, consider the thin homogeneous representations of the quiver $\tilde{A}_{1,2}$, they are of the form

$$S(\lambda) = \quad \begin{array}{ccc} & k \xrightarrow{\ \lambda\ } k & \\ {}^{1}\!\!\searrow & & \nearrow^{1} \\ & k & \end{array}$$

and are pairwise non-isomorphic, for $\lambda \in k$. Note that we have $\sigma S(\lambda) \simeq S(-\lambda)$, thus, if the characteristic of k is different from 2, then $\sigma S(\lambda)$ is not isomorphic to $S(\lambda)$. Note that in this case we have $\mathrm{Ext}^1(S(\lambda), C^+ S(\lambda)) = 0$, whereas, of course, $\text{Ext}^1(S(\lambda), D \operatorname{Tr} S(\lambda)) \neq 0$.

For some indecomposable representations V, the representations C^+V and $D \operatorname{Tr} V$ may be isomorphic, in particular this is true in case V and therefore $D \operatorname{Tr} V$ is uniquely determined by the dimension vector. Actually, the Coxeter functor has been used mainly for those modules where it coincides with $D \operatorname{Tr}$. It is the Auslander-Reiten construction $D \operatorname{Tr}$ which plays the decisive role. Many of the investigations carried out before 1975 were based on an explicit or at least implicit use of Coxeter functors (we mention the topics 4, 6, 7). The interpretation of C^+ as a sort of substitute for $D \operatorname{Tr}$ shows that the appearance of the Auslander-Reiten paper in 1975 should be considered as a natural closing: The Auslander-Reiten theory serves in many ways as the proper setting for the previous investigations.

As we have noted in [1], it is very intriguing to see the interplay between the homological and categorical methods of the Auslander-Reiten theory and the combinatorial approach developed in Moscow, Zürich and Kiev. A convenient way in order to determine the module category of suitable algebras is to work inductively, using one point extensions and the representation theory of posets. At any step, the best way to obtain the poset needed is to determine the corresponding part of an Auslander-Reiten quiver. Note that the meshes of the Auslander-Reiten quiver explain well the structure of the posets which arise. The concept of a hammock interrelates these categorical and combinatorical features. By now, almost all combinatorial methods in representation theory have found a homological interpretation, and it is the Auslander-Reiten theory which usually is invoked to do so. The different approaches which were established during the period 1968 to 1975 now have to be seen as parts of a unified theory.

Final remarks.

We have selected those topics treated between 1968 and 1975 which are still of utmost interest and which were brought to a satisfactory state then. There had been attempts to deal with other open questions such as the second Brauer-Thrall conjecture and the Gelfand problem, but they were unsuccessfull at that time. We admit that we have bypassed two major treatises which appeared during the period in question. First, there is the work of Auslander and Bridger [AB] from 1969, but it has to be considered as a general study in homological algebra not being confined to finite dimensional algebras. Second, Tachikawa's investigation [T] dealing with QF-1 and QF-3 rings was published in 1973, its main interest however lies in generalizing results known for finite dimensional algebras to more general classes of rings. As a main principle, the discussion has been concentrated on the core of the subject. We have left aside the relationship to ring theory, to model theory and to algebraic geometry, but also possible applications.

Some readers may miss a stronger insistance on the functorial approach to representation theory. The developments of these years can be considered also in the general framework of functor categories (after all, any module category is a functor category), for a short outline let me refer to section 9 of [2] which deals with the same period. However, in my opinion, the use of functor categories was not one of the decisive themes during the period in question. Note that the basic references such as Gabriel's thesis and Auslander's La Jolla paper on finitely presented functors were published before that time and the main emphasis during the period 1968–1975 was shifted towards the working with concrete algebras, modules, and exact sequences. Of course, the homological and categorical knowledge which was assembled before was kept in mind and was properly used, but there was a strong reluctance to engage in general nonsense.

Main References

[A1] M. Auslander: Representation Dimension of Artin Algebras. Queen Mary College. Math. Notes. (1971)

[A2] M. Auslander: Representation theory of artin algebras. II. Comm. Algebra 1 (1974), 269-310.

[AB] M. Auslander, M. Bridger: Stable module theory. Memoirs Amer. Math. Soc. 94. Providence (1969).

[AR1] M. Auslander, I. Reiten: Representation theory of artin algebras. III. Comm. Algebra 3 (1975), 239–294; IV. Comm. Algebra 5 (1977), 443-518.

[AR2] M. Auslander, I. Reiten: Stable equivalence of dualizing R-varietes. I. Advances in Math. 12 (1974), 306–366; II. Advances in Math. 17 (1975), 93–121; III. Advances in Math. 17 (1975), 122–142; IV. Advances in Math. 17 (1975) 143–166.

[BGP] I.N. Bernstein, I.M. Gelfand, V.A. Ponomarev: Coxeter functors and Gabriel's theorem. Uspechi Mat. Nauk. 28 (1973); translation: Russian Math. Surveys 28 (1973), 17–32.

[B1] S. Brenner: Some modules with nearly described endomorphism rings. J. Algebra 23 (1972), 250-262.

[B2] S. Brenner: Decomposition properties of some small diagrams of modules. Symposia Math. XIII, Ist. Naz. Alta Mat. (1974) 127–141.

[C] A.L.S. Corner: Endomorphism algebras of large modules with distinguished submodules. J. Alg. 11 (1969), 155-185.

[DR1] V. Dlab, C.M. Ringel: On algebras of finite representation type. J. Algebra 33 (1975), 306–394.

[DR2] V. Dlab, C.M. Ringel: Représentations des graphes valués. C.R. Acad. Sc. Paris A 278 (1975), 537–540.

[DF] P. Donovan, M.R. Freislich: The representation theory of finite graphs and associated algebras. Carleton Lecture Notes 5, Ottawa (1973).

[D] Yu.A. Drozd: Coxeter transformations and representations of partially ordered sets. Funkc. Anal. i Pril. 8 (1974), 34–42. Translation: Funct. Anal. and Appl. 8 (1974), 219–225.

[G1] P. Gabriel: Unzerlegbare Darstellungen I. Manuscripta Math. 6 (1972), 71–103.

[G2] P. Gabriel: Indecomposable representations II. Symposia Math. Ist. Naz. Alta Mat. 11 (1973), 81-104.

[G3] P. Gabriel: Représentations indécomposables des ensembles ordonnés. Séminaire P. Dubreil, Paris (1972–73), 301–304.

[GP1] I.M. Gelfand, V.A. Ponomarev: Indecomposable representations of the Lorentz group. Russian Math. Surveys 23 (1968), 1-58.

[GP2] I.M. Gelfand, V.A. Ponomarev: Problems of linear algebra and classification of quadruples in a finite-dimensional vector space. Coll. Math. Soc. Bolyai 5, Tihany (1970), 163–237.

[K1] M.M. Kleiner: Partially ordered sets of finite type. Zap. Naučn. Sem. LOMI 28 (1972), 32–41. Translation: J. Soviet Math. 23 (1975), 607–615.

[K2] M.M. Kleiner: On exact representations of partially ordered sets of finite type. Zap. Naučn. Sem. LOMI 28 (1972), 42–60. Translation: J. Soviet Math. 23 (1975), 616–628.

[N1] L.A. Nazarova: Representations of quivers of infinite type. Izv. Akad. Nauk SSSR, Ser. Mat. 37 (1973), 752–791.

[N2] L.A. Nazarova: Partially ordered sets of infinite type. Izv. Akad. Nauk SSSR, Ser. Mat. 39 (1975), 911–938.

[NR] L.A. Nazarova, A.V. Roiter: Representations of partially ordered sets. Zap. Naučn. Sem. LOMI 28 (1972), 5–31. Translation: J. Soviet Math. 23 (1975), 585–607.

[R1] C.M. Ringel: The indecomposable representations of the dihedral 2-groups. Math. Ann. 214 (1975), 19–34.

[R2] C.M. Ringel: The representation type of local algebras. Proc. ICRA 1974. Springer LNM 488 (1975), 282–305.

[RT] C.M. Ringel, H. Tachikawa: QF–3 rings. J. Reine Angew. Math. 272 (1975), 49–72.

[R] A.V. Roiter: The unboundedness of the dimension of the indecomposable representations of algebras that have an infinite number of indecomposable representations. Izv. Acad. Nauk SSSR, Ser. Math. 32 (1968), 1275–1282. Translation: Math. USSR, Izv. 2 (1968), 1223–1230.

[T] H. Tachikawa: Quasi-Frobenius rings and generalizations. Springer LNM 351 (1973)

Additional References

[1] C.M. Ringel: Tame algebras. In: Representation Theory I. Springer LNM 831 (1980), 137–287.

[2] C.M. Ringel: Maurice Auslander and the Representation Theory of Artin Algebras. CMS Conference Proceedings 18 (1996), 4-12.

FAKULTÄT FÜR MATHEMATIK
UNIVERSITÄT BIELEFELD
POBox 100131
D-33501 BIELEFELD
GERMANY
E-mail address: ringel@mathematik.uni-bielefeld.de

ALGEBRAIC GEOMETRY OVER $\overline{\mathbb{Q}}$

LUCIEN SZPIRO

In memory of Maurice Auslander

In 1978 I gave a talk at the ENS in Paris. The title was "Faisceaux arithmétiques coherents". My goal was to introduce the mathematical public to what I called "Arakelov theory". To justify this introduction I explained what could be done with the idea of Arakelov and Parshin: "Put metrics at infinity on vector bundles and you will have a geometric intuition of compact varieties to help you". I also explained that my seminar [Sz 1] written in geometric language, could be considered as a book of conjectures once one knew the translation of effective divisor, Kodaira-vanishing theorem, bounded families, Hodge index etc... Needless to say I did not raise enthusiasm at this point!

I present here the work that has been done on this program.

1. Faisceaux arithmétiques cohérents.

1-1 Heights.

Let K be a number field. The local-global equality defining the height of a point $\mathbf{x} \in \mathbb{P}^n(K)$ is: (with $L := \mathcal{O}(1)$)

$$(\star)\, h(\mathbf{x}) = \frac{1}{K : \mathbb{Q}} \log \frac{\mathrm{vol}(\mathcal{O}_K)}{\mathrm{vol}(L/E_\mathbf{x})} = \frac{1}{K : \mathbb{R}} \log \frac{\prod_v \sup_i |x_i|v}{N(\sum x_i \mathcal{O}_K)}$$

where $E_\mathbf{x}$ is the section of $\mathbb{P}^n_{\mathcal{O}_K} \to \mathrm{Spec}\,\mathcal{O}_K$ corresponding to the point $\mathbf{x}$. This formula teaches us many things:

(1) It is a **Riemann-Roch** theorem in dimension one analogous to $\chi(L) = \deg L - g + 1$ on a curve.

(2) **The height is the "intersection"** of a scheme of dimension 1 with a cycle of codimension 1. The fundamental theorem on heights is typical of the type of results one is able to get about $\overline{\mathbb{Q}}$:

Theorem 1 (Northcott's theorem).

Given $d \in \mathbb{N}$ and $A \in \mathbb{R}_+$ the set

$$\{\mathbf{x} \in \mathbb{P}^n(\overline{\mathbb{R}}) | h(\mathbf{x}) \le A; \deg[K(\mathbf{x}) : \mathbb{Q}] \le d\}$$

is finite.

As a corollary, once one knows $\hat{h}$ (the **Neron-Tate height** on an abelian variety A) one gets the finiteness of the torsion of $A(K)$ because $(\hat{h}(P) = 0 \Leftrightarrow P$ has torsion). One also gets: [weak (Mordell-Weil) $\Rightarrow$ (**Mordell-Weil** ($A(K)$ is of finite type over $\mathbb{Z}$)].

1-2 Basic theorems of algebraic number theory

One can now prove, in the language of metrized line bundles on $\mathrm{Spec}\mathcal{O}_K$, the following classical results: $cl(\mathcal{O}/K)$ is finite, $|d_K| > 1$, Dirichlet units theorem, and that for a fixed d_K there is only a finite number of K possible. In fact this last statement is not sufficient for the purposes of this paper. One needs – and one proves – the following

Theorem 2.

Given $n \in \mathbb{N}$ and a finite set of primes $p_1 - p_r$ in $\mathcal{O}_K$. The set of number fields L such that $[L : K] \leq n$ and $\mathcal{O}_L$ is ramified over $\mathcal{O}_K$ only over the $p_i s$ is finite.

The proof of this **is not** "arakelovian". One has to bound the wild ramification (something false in characteristic $p > 0$). This is typical: arakelovian methods give the finiteness of certain objects in $\bar{\mathbb{Q}}$, then the arithmetic allows one to work over a given number field. In particular theorem 2 is needed to prove the weak Mordell-Weil theorem.

1-3 The paper of Arakelov

In 1972 Arakelov introduced an intersection theory on arithmetic surfaces with the following properties:
(i) The theory satisfies an **Adjunction** formula (with a "grain de sel" when the divisor is not a section).
(ii) The theory extends the **Neron-Tate** pairing on divisors of degree zero.
    One should remark at this point that an arithmetic surface $X \to \mathrm{Spec}\mathcal{O}_K$ is analogous to a surface fibered over a compact curve but with **no fixed part in the Jacobian**. In fact the Neron-Tate height is zero on the fixed part and not only on torsion points.
(iii) The theory of **admissible metrics** he introduced are of deep arithmetic content. Two cases have been worked on extensively :
    a) the case of **elliptic curves** has been worked out quite completely in [Sz 2] after a start in [F 1].
    b) the **self intersection** $(\omega_{X_{\mathcal{O}_K}} \cdot \omega_{X_{\mathcal{O}_K}})$ is a new invariant for curves of genus at least 2. It is exploited below (2-1).

1-4 Cohomology of coherent sheaves

Two versions of the **Riemann-Roch** theorem computing the volume for the Quillen metric of $Rf_\star E$ for a metrized line bundle on a generically smooth $X \to \mathrm{Spec}\mathcal{O}_K$ has been proposed by Gillet-Soulé and Faltings. One should note that it is not clear that the two versions of the Riemann-Roch coincide. The most useful part of a Riemann-Roch theorem is in its leading term:

Theorem 3 (The arithmetic Hilbert-Samuel theorem for generically smooth varieties).

Let L be a positively metrized ample line bundle on $f : X \to Spec\mathcal{O}_K$,

$$if\, n \gg 0\, then\, -\log vol_{L^2}(f_* L^{\otimes n}) = \frac{n^{d+1}}{(d+1)!}(L \cdots L) + o(n^{d+1})$$

This can be used as an alternative definition of $(L \cdots L)$, $(d+1$ times). (d is the dimension of the generic fiber). The following corollaries were indicated in my talk in '78:

Corollary 1 (existence theorem for small sections).

If $(L \cdots L) > 0$ and $n \gg 0$ there exists an $s \in f_* L^{\otimes n}$ such that

$$\|s\|_{L^2_\sigma} \leq 1$$

for every place σ at infinity.

Proof. Apply Minkowski's famous theorem on lattice points.

Corollary 2.

If H is numerically ample and $(L_{/H} \cdot L_{/H} \cdots L_{/H})$ (d times) is zero then $(L \cdot L \cdots L) \leq 0$ ($d+1$ times). (The **index theorem** is most striking when $d = 1 : (L \cdot H = 0$ implies $(L \cdot L) \leq 0))$

1-5 More applications of Corollary 1.

The existence of small sections is crucial in many cases; we list a few below: (One should note the paper of **Abbes and Bouche** which gives in less than 30 pages a self contained proof of theorem 3 ([A-B])) (cf. also [R]).

(i) $(\omega_{X\mathcal{O}_K} \cdot \omega_{X\mathcal{O}_K}) \geq 0$ $g \geq 1$ (Faltings [F 1])
(ii) A second proof of the Mordell conjecture (**Vojta** [V])
(iii) **Miyaoka's** $c_1^2 \leq 4c_2$ for stables vector bundles on curves $g \geq 2$
(iv) My proof of Bogomolov's conjecture generalizing Raynaud's theorem [Sz 2]
(v) **Shouwu Zhang's** Nakai Moisheson's criterion of ampleness [Zh 1]

1-6 The degree of a subvariety is the self-intersection.

Theorem 3′(arithmetic Hilbert Samuel for any variety). Let L be a positively metrized ample line bundle on $f : X \to Spec\mathcal{O}_K$, then if $n \gg 0$ one has

$$-\log vol_{L^2} f_* L^{\otimes n} = \frac{n^{d+1}}{(d+1)!}(L \cdots L) + o(n^{d+1})$$

when the L^2 norm is computed on the smooth locus of the reduced variety.

This is deduced from theorem 3 by S. Zhang in [Zh 2] using resolution of singularities).

The next useful result is:

Theorem 4.
The cut out of a small hyperplane section gives a subvariety of small degree.

(This is made precise in ([F 2]). This statement with corollary 1 of theorem 3 is what is needed of the theory in the proof by Faltings of his famous "product theorem". Then, with that in hand, one is able to prove Lang's conjecture for abelian varieties:

Theorem 5. *Let A be an Abelian variety over a number field K and let X be a subvariety of A. Then $X(K)$ is contained in a finite union of $B_i(K)$, where the B_i are translates of proper sub-abelian varieties of A.*

1-7 Grothendieck-Riemann-Roch?

For a morphism $X \xrightarrow{f} Y$ over $\mathrm{Spec}\mathcal{O}_K$ a few cases have been done:
(i) Max Noether's formula $12X(\mathcal{O}_X) = c_1 + 4c_2$ (Faltings [F 1])
(ii) Max Noether's formula on $\mathcal{M}_g$ (L.Moret-Bailly [M-B 1])
(iii) Functional equation of θ functions (L. Moret-Bailly [M-B 2])
(iv) $c_1(Rf_\star)$ for f a local complete intersection (**Lin Weng** [W])
(v) $c_1(Rf_\star)$ for Macaulay schemes (R. Elkik ([E 1], [E 2])

2. A la recherche de petits points: Numerical properties of the relative dualizing.

To try to get (conjecturally so far) effectivity statements I proposed in [Sz 1] to look at small points. The reasons were:
a) the **Parshin-Kodaira construction** tells us that one is interested in finding an upper bound for the quantity $(\omega_{X/\mathcal{O}_K})^2$,
b) the following lemma that I had proved:

Lemma 5.
$$(\omega_{X/\mathcal{O}_K} \cdot \omega_{X/\mathcal{O}_K}) \leq (-E_P^2)(2g - 2)(2g)$$
where X_K is a curve of genus $g \geq L$ and E_P the section of $X \to \mathrm{Spec}\mathcal{O}_K$ given by a rational point $P \in X_K(K)$.

2-1 $\omega_{X/\mathcal{O}_K}$ is big (i.e. points are not too small).

Theorem 6 (discreteness of algebraic points).
Let X be an arithmetic surface of genus $g \geq 2$ and $X_K \hookrightarrow A$ an embedding of X_K in a polarised abelian variety, then there exists an $\varepsilon > 0$ such that $\{P \in X_K(\overline{\mathbb{Q}}) | \hat{h}(P) \leq \varepsilon\}$ is finite except perhaps when P "divides" $\omega_{X/\mathcal{O}_K}$. When X is smooth over $\mathcal{O}_K$ then $\omega_{X/\mathcal{O}_K}^2 > 0$ is equivalent to the full finiteness statement above.

This result, which generalized **Raynaud's** famous theorem on torsion points, was conjectured by **Bogomolov**. The exceptional case is very interesting. It gives an arithmetic meaning to the Arakelov invariant $\omega_{X/\mathcal{O}_K}^2$.

To prove that $\omega^2_{X/\mathcal{O}_K} = 0$ implies that there is an infinite sequence of points $x_n \in X(\bar{\mathbb{Q}})$ with $\hat{h}(x_n) \to 0$, I needed the fact that the linear system $f_*\omega^{\otimes n}_{X/\mathcal{O}_K}$ has no fixed part. This has been proved by Kim ([K]). A more general theorem has been obtained by **S. Zhang [Z 1]**.

Theorem 7 (Nakai Moisheson theorem for arithmetic surface).
If L is numerically ample (i.e. $(L \cdot L > 0 \; L \cdot D > 0$ for any effective D) then $L^{\otimes n}$ is generated by its sections that are smaller than one in the L^2 norm, provided $n \gg 0$.

We have recently, with Ullmo and Zhang, proved that the generalized Bogomolov conjecture is equivalent to the equirepartition of small points [SUZ].

2-2 When is $(\omega_{X/\mathcal{O}_K} \cdot \omega_{X/\mathcal{O}_K}) > 0$?

S. Zhang [Zh 2] proved the following:

Theorem 8.
*If $X \longrightarrow Spec\mathcal{O}_K$ is semi-stable and **not smooth** of genus $g \geq 2$ then $(\omega_{X/\mathcal{O}_K} \cdot \omega_{X/\mathcal{O}_K}) > 0$.*

In the smooth case not everything is known. The following authors have given examples of $(\omega_{X/\mathcal{O}_K} \cdot \omega_{X/\mathcal{O}_K}) > 0$ for smooth fibrations. (Burnol), (S. Zhang), (Mestre, Bost, Moret-Bailly).

2-3 $\omega_{X/\mathcal{O}_K}$ must be small

To prove such a statement I have in [Sz 1] a program covering two points
(i) - Kodaira vanishing
(ii) - Kodaira Spencer class.
In fact (i) was essentially solved by **Miyaoka** 10 years ago:

Theorem 9 (Miyaoka).
If E is a stable vector bundle on an arithmetic surface then $c_1^2 \leq 4c_2$.

This has been written up by **Moriwaki**. The vanishing theorem is then deduced by the Mumford-Reider method. C. Soulé has published the proof of the following [S 1]:

Corollary 10.
If L is a numerically ample line bundle on an arithmetic surface and $s \in R^1 f_* L^{\otimes -1}$, then $\|s\|_{L^2} \geq e \; (\log e = 1)$.

2-4 The instrinsic small points conjecture.

I have made numerous variations on the "small points" I had obtained in the geometric case [Sz 1]. The following is a version I rather like: Let X_K be a curve of genus $g \geq 2$ over a number field K_0. Then there exist constants $A(n)$ and $B(n)$ depending only on the curve over K_0 and on the integer n such that : if $(K : \mathbb{Q}) \leq n$, for every $P \in X(K) \; \exists \; P' \in X(\bar{\mathbb{Q}})$ such that

(i) $E^2_{P'} \leq [K(P') : \mathbb{Q}](A(n)\log D_K + B(n))$

(ii) $(E_P \cdot E_{P'}) \leq [KP, P') : \mathbb{Q}](A(n)\log D_K + B(n))$.

This implies a strong effective Mordell conjecture which, if proved for **only one curve**, would imply the (a,b,c) conjecture (or the conjecture of discriminant for elliptic curves) — a result of **L. Moret-Bailly** and myself. In this direction note the theorem of **E. Ullmo**:

Theorem 11.

For every point $P \in X(K)$ there exists a $Q \in X(\overline{\mathbb{Q}})$ such that $(E_P \cdot E_Q)_{\text{finite}} = \varnothing$ with height (Q) bounded.

Of course there remains the problem of finding an upper bound for $(E_P \cdot E_Q)_\infty$!!

3. What is our subject?

3-1 Integral models or adelic studies

The difficulties of looking at the geometrical model, integral over $\text{Spec}\mathcal{O}_K$, has lead many authors to try to go back to the Weil-adelic point of view. In this direction S. Zhang [Zh 2], Rumely, Soulé-Gillet-Bloch have developed an adelic Arakelov theory. In [Zh 2] S. Zhang uses his theory to prove Theorem 8.

3-2 What metrics to choose

Each metric has its problem

i) Faltings put $\int \omega \wedge \overline{\omega}$ on $\overset{\max}{\wedge} \Omega^1_A$ for A an abelian variety with the success we know. It lead him to ask himself the purely arithmetic question of how to evaluate the discriminant of the kernel of an isogeny. This is a good example of the following philosophy:

α) determine a statement on $\overline{\mathbb{Q}}$ using a metric (here prove that the modular height has logarithmic singularities and satisfies Northcott's theorem)

β) Then use arithmetic (here Raynaud's $(p, \dots, p)$ theorem and Grothendieck-Illusie's "Barsotti-Tate tronqués")

ii) S. Zhang these days seems to prefer the Poincaré metric to Arakelov's original one for it extends over the closure of the Deligne-Mumford compartification of $\mathcal{M}_g$

iii) In defense of admissible Arakelov metrics I will quote:

α) it extends Neron-Tate pairing (hence Theorem 6)

β) $\omega^2_{X/\mathcal{O}_K}$ is a height with log singularities on $\mathcal{M}_g$ as the work of **Jorgenson** on Faltings' δ function shows

γ) I proved in [Sz 2] that $12 \deg_{\text{Arakelov}}(\omega) = \log D_{\min}$ for an elliptic curve

iv) Soulé, Gillet, Bismut have choosen the Quillen metric for it is the one that gives them a Riemann-Roch theorem.

3-3 Galois action

It is my conviction that to really prove statements about a number field (or, better, on $\mathbb{Q}$) and not on $\overline{\mathbb{Q}}$ (as an inductive limit) one has to put not only metrics at infinity but to take into account the Galois action.

Faltings [F 3] is, as I note above, a perfect example.

To give a chance to the opposite opinion I will note two things:

α) My conjecture 2-4 about intrinsic small points is purely on $\overline{\mathbb{Q}}$ (except that the discriminant D_K is there).

β) E. Ullmo has proved the following:

Theorem 12.

Let X be an semi-state elliptic curve, $d = \frac{1}{12} log(D_{\min})$ then for every $\varepsilon > 0$ there exists only a finite number of **torsion points** P in $X(\overline{\mathbb{Q}})$ such that $-\phi_P^2/[K(P):K] \leq d - \varepsilon$. Here ϕ_P is the $\mathbb{Q}$-divisor that makes $E_P - E_0 + \phi_P$ purely of degree zero.

This is a clever corollary of [Zh 1] and [Sz 2]. If $d > 0$ (i.e. if X **does not** have good reduction). Theorem 12 implies that **the set of torsion points which always specialize inside the connected component of the Néron model is finite over** $\overline{\mathbb{Q}}$. This could be obtained by Galois considerations using Serre's theorem on the irreducibility of the Galois image in $Gl_2(\mathbb{F})$ for $l \gg 0$. Here the proof is global and arakelovian.

Added in September '96:

The Bogomolov conjecture for curves was proved in June 96 by E. Ullmo (to appear). Using his method (a smart double use of the equidistribution theorem [SUZ]) S. Zhang proved the general case in July 96 (to appear). In particular E. Ullmo proves in his paper: $(\omega_{X/\mathcal{O}_k})^2 > 0$ for every arithmetic surface of genus greater than one.

For EU product safety concerns, contact us at Calle de José Abascal, 56–1°,
28003 Madrid, Spain or eugpsr@cambridge.org.

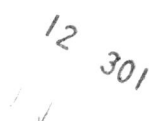

 www.ingramcontent.com/pod-product-compliance
Ingram Content Group UK Ltd.
Pitfield, Milton Keynes, MK11 3LW, UK
UKHW012315141225
465965UK00001B/80